Hybrid Renewable Energy Systems

Scrivener Publishing
100 Cummings Center, Suite 541J
Beverly, MA 01915-6106

Advances in Renewable Energy

Series Editor: Umakanta Sahoo

Scope: The global warming phenomenon as a significant sustainability issue is gaining worldwide support for development of renewable energy technologies. Advance in Renewable Energy covers the recent development on solar energy technologies (Solar Photovoltaic & Solar Thermal) for various applications like power generation, process heat etc., Design development of hybrid renewable energy systems i.e. wind-photovoltaic, solar thermal with biomass, bagasse, bio gas and New development of energy storage for continuous generation of power, various industrial process heat applications, cooling, desalination, and other areas.

Submission to the series:
Contact Publisher Phil Carmical
pcarmical@scrivenerpublishing.com
(512)203-2236

Publishers at Scrivener
Martin Scrivener (martin@scrivenerpublishing.com)
Phillip Carmical (pcarmical@scrivenerpublishing.com)

Hybrid Renewable Energy Systems

Edited by

Umakanta Sahoo

Scrivener Publishing

WILEY

This edition first published 2021 by John Wiley & Sons, Inc., 111 River Street, Hoboken, NJ 07030, USA
and Scrivener Publishing LLC, 100 Cummings Center, Suite 541J, Beverly, MA 01915, USA

For more information about Scrivener publications please visit www.scrivenerpublishing.com.

Wiley Global Headquarters
111 River Street, Hoboken, NJ 07030, USA

For details of our global editorial offices, customer services, and more information about Wiley products visit us at www.wiley.com.

Library of Congress Cataloging-in-Publication Data

ISBN 9781119555575

Cover image: Green Energy Source Icons | Pop Nukoonrat, dreamstime.com
Cover design by Kris Hackerott

Set in size of 11pt and Minion Pro by Manila Typesetting Company, Makati, Philippines

Contents

1

Resource Assessment and Implementation of Hybrid Renewable Energy Systems for Food Preservation in Agro-Tropical Areas: A Techno-Economic Approach

M. Edwin[1]*, M. Saranya Nair[2] and S. Joseph Sekhar[3]

[1]*Department of Mechanical Engineering, University College of Engineering, Nagercoil, Anna University Constituent College, Nagercoil, India*
[2]*School of Electronics Engineering, Vellore Institute of Technology, Chennai Campus, Chennai, India*
[3]*Mechanical Section, Department of Engineering, Shinas College of Technology, Shinas, Sultanate of Oman*

Abstract

In tropical countries, approximately 75% of people are involved in the production of dairy and agricultural goods. About 45% of agricultural produce loses its value in rural areas due to the lack of adequate storage and transport facilities. Therefore, cooling systems that work with locally available renewable energy resources are essential to preserving the perishable agricultural products in isolated tropical regions. By implementing a suitable vapour absorption cooling system, the heat energy from the available renewable energy resources can be used to satisfy the cooling demand. This chapter examines the techno-economic viability of the number of configurations of producer gas, biogas and gobar gas energy sources that operate a chilling system. The analysis has been performed with the use of MATLAB software, and the suitable hybrid energy systems have been predicted for serving a community in tropical regions. The results indicate that producer gas/biogas/gobar gas combinations in hybrid renewable energy-operated 35 kW vapour absorption cooling system shows the minimum operating cost (INR 0.81–0.88 million), simple payback time (2.8–3.2 years) and life cycle cost (INR

**Corresponding author*: edwinme1980@gmail.com

Umakanta Sahoo (ed.) Hybrid Renewable Energy Systems, (1–36) © 2021 Scrivener Publishing LLC

44.5–48.4 million), and effectively achieves a positive net present cost and the maximum overall efficiency (0.19–0.24).

Keywords: Hybrid energy system, bioenergy, absorption chiller, payback time, net present cost, life cycle cost

1.1 Introduction

Energy demand is increasing at an unprecedented rate all over the world. A major part of the world's current energy generation is from fossil fuels, resulting in the continuous depletion of this resource. In tropical countries, approximately 75% of the population is engaged in the production of milk and agro products. About 25-45% of agricultural produce loses its value in rural areas due to the lack of adequate storage and transport facilities. At the local village level, this spoilage can be avoided by providing suitable cooling systems for short-term preservation [1]. Integration of available renewable energy sources for energy systems in remote tropical areas has become an area of current interest in the renewable energy sector. A major part of India's population lives in rural areas and many such villages in those areas are disconnected from the grid connectivity. The extension of grid power to these villages may not be feasible due to real-time situations; however, a stand-alone hybrid renewable energy system can be implemented to overcome such issues [2]. Cleaner energy in the electricity grid is necessary to mitigate the rising greenhouse gases. Due to the fluctuating nature of renewable energy, its outputs are subject to variation and hence it is important to plan as much buffer as possible in the portion of renewable energy in any energy system [3]. Several works have been carried out regarding the feasibility of hybrid renewable energy-operated refrigeration systems, and its implementation issues, capacity, efficiency, optimisation and combinations with other renewable energy technologies. Normally hybrid energy is considered the working of primary renewable energy systems with conventional energy sources. The renewable energy hybrid system will be cost-competitive for rural areas, instead of stand-alone systems [4]. Moreover, their optimum combination would reduce environmental pollution, capital and operating expenses [5]. The combined impacts of several variables, including expense, productivity, social impact, acceptance, reliability and demand are the essential parameters for the implementation of a hybrid energy system [6]. Synergistic advantages can be obtained by integrating multiple forms of energy resources in the generation of heat and electricity. When settling on the viability of a given arrangement, close attention must be paid to many crucial considerations such as the origin of the source, access to different renewable energy sources, co-location of the sources, handling issues in the energy source, etc. While deciding the feasibility of a hybrid system, the possibility of combining the different energy sources in a particular area of the application is more essential [7].

It is noted that the suitable mixtures of solar, biomass and biogas system would improve the techno-economic and socio-economic status in remote areas. The integration of a biomass and biogas energy system can reduce the accessibility issues of the solar energy conversion system, and reduce the implementation cost of the hybrid energy system. This shows that the integration of two or more forms of energy systems can balance the strength and weakness of each other [8]. The proper integration of available bio-energy sources can overcome the solar dependency issues caused by the peculiar climatic and geographical conditions of remote places [9]. The integration of solar and biogas to supply heat to a vapour absorption chiller is economically viable and the excess biogas in this process can be used to operate the electrical accessories, including pumps, in the cooling system [10]. This combination of these two types of renewable energy sources gives the COP of chiller and overall system as 0.7 and 0.11, respectively [11]. The study also shows that the solar-biomass hybrid system is capable to act with better precision and efficiency than traditional systems with more than 65% savings in operation time [12]. Even though the hybrid solar-biomass system has environmental benefits, a long payback period is reported as the major issue [13]. Biomass gasifier is integrated with the wind energy conversion system to mitigate the prediction uncertainties in stand-alone wind power systems [14]. The mixed linear programming model for the demand side management has been used to efficiently use renewable energy sources such as solar, biomass, biogas and hydro with a diesel generator to reduce the system's capital cost [15].

In a conventional power plant, the integration of renewable energy sources can lower 30% of the net present cost and 16t of CO_2 emission [16]. The energy cost biomass/hydro/solar based hybrid energy system has been reported as INR 7.5/kWh, which is lower than that of an existing hybrid system working with hydro and diesel energy sources [17]. A similar reduction in electricity cost is observed when a hybrid power system works with fuel cells, which leads to the low operating cost and favourable payback time [18].

A positive return of investment and 95% reduction in carbon emissions can be obtained in a wind/solar/diesel hybrid system besides a positive net present value, good return on investment and a Levelized cost of 0.387 $/kWh [19]. It is found that the wind-solar hybrid energy system gives an optimal net present cost and energy cost of $676,500 and 0.274 $/kWh, respectively. Moreover, it prevents 6,100 tons of CO_2 to enter the air and also saves $121,600 over 25 years of its lifetime [20]. The energy cost of INR10.18/kWh is reported for a solar/wind/bioenergy-based hybrid power generation for a remote area in the southern part of India. When the bioenergy sources are integrated with solar and wind energy conversion systems, the capital investment cost of the system can be minimised [21]. The non-uniform availability of these energy sources throughout a year leads to fluctuation in the power output, and such issues can be tackled by introducing an appropriate hybrid system with one-hundred present renewable energy sources [22, 23]. Selection of wind power among all renewables may not be the best economic decision, because biomass emerges as an essential attribute to an efficient hybrid system [24]. As a renewable energy source, biomass waste will dramatically increase the techno-economic efficiency of local hybrid renewable energy systems and support the government agencies to overcome the issues related to dumping of waste and its impacts on soil quality. Further, the reliability of such systems is proved by the sensitivity analysis on its critical factors [25]. Biogas/biomass/solar/wind/fuel-cell hybrid energy system can supply energy at a low cost of $0.163 per kWh [26]. It is observed that the biogas conversion contributes up to a 65% share of electricity production, thereby providing a reliable community-scale energy generation capacity along with reducing the disposal costs of local solid waste. The net present cost of biomass integrated hybrid system shows its advantage during the operation [27]. However, sufficient studies are needed to optimise the contributions of individual energy sources in a hybrid system to improve the overall efficiency, cost advantages, reliability and the impact of economic indicators to match the capacity [28].

It is seen from the review of the literature that most of the works on the hybrid renewable energy systems focus on power generation. The implementation of the hybrid-energy concept in thermal applications is also essential to further widen the application of renewable energy sources in remote places. Therefore, this study is focused on the assessment of renewable energy sources and the thermal energy requirements for food preservation in isolated places and identify a suitable combination of bioenergy sources for maximum system efficiency, economic advantage and environmental benefits. Since the total energy needed for a particular application is taken from renewable energy sources, the mitigation of environmental issues is expected.

1.1.1 Objectives

Most of the food preservation technologies prevailing today use electricity from either grid or fossil fuel–operated engines to run the vapour compression cooling systems. A few solar power–operated systems are also installed in some places. These systems have issues either with environmental impacts or high initial investment. Many hurdles are experienced in using multiple renewable energy sources since the vapour compression systems need electrical power as input energy. However, the advantage of renewable energy sources can be utilized by operating a vapour absorption system from the low-grade heat which is taken from the renewable energy sources with high conversion efficiency. Therefore, the bioenergy conversion technology was considered for the efficient installation of an absorption cooling system to conserve the agricultural produce of some remote locations in the southern part of India, where grid connectivity and transport facilities are limited. To accomplish this task the major objectives are defined as follows.

- To assess the cooling needs and the potential of renewable energy sources based on the data collected in the field survey.
- To find the performance parameters of the selected vapour absorption chilling system working with different combinations of producer gas, biogas and gobar gas in an agro-tropical area.
- To predict the economic parameters for the implementation of the proposed system.
- To identify the suitable combination of renewable energy sources for food preservation in the selected area, with optimum overall efficiency, financial advantages and risk parameters.

1.2 Materials and Methods

The energy-efficient and eco-friendly preservation of agricultural products including milk at the production area itself can improve the quality and reduce the spoilage perishable produce. This can also contribute to the world's growth because its essential connection and growth opportunities enhance the two economic pillars, agriculture and manufacturing. Since agriculturally based activities are clustered mainly in a rural area, proper care should be given to those areas in applying renewable energy technologies with economic benefits. Therefore, some villages of such area in the Western Ghats of India have been taken for the study and the available energy sources, and their demands for preservation needs are studied to identify the appropriate mix of energy sources to run a vapour absorption system to meet the cooling need.

In this section, the procedure used to identify the critical factors considered to survey local households, the energy needed to run a vapour absorption cooling system to meet the preservation needs and the possible mix of energy sources are discussed. A load/demand analysis was also performed to identify the appropriate capacity of the cooling system.

1.2.1 Resource Assessment

1.2.1.1 Definition of the Study Region

A cluster of villages in south Tamil Nadu, which was an industrial-backward district in India, was selected for the study. Relevant data was gathered through a comprehensive field-level survey of the information available from census and other sources. A schedule for house listing was developed and distributed to all households in the study area. The house listing plan included all households in the villages selected and specific details, such as the number of families, profession, the scale of the land-holdings in operation, etc. The photographs taken from the study region are shown in Figure 1.1. A systematic procedure for primary field survey about rural energy use is employed by taking into account the most important criteria affecting the trend and strength of energy usage.

1.2.1.2 Field Survey from Households

Based on the past research, samples have been collected regarding the domestic energy consumption, energy consumption in agriculture, milk

Figure 1.1 Photographical view of the study region.

production and preservation needs, the economic viability of technologies in the rural systems and impact of technology on rural systems. Households were mentioned in alphabetical order in the study region for the family head numbers, including 10–15 households chosen by systematic random sampling process from the study region. Based on this, 60 sample houses in each area are chosen, and the comprehensive household-level analysis is performed.

Field studies on household and direct interview approach in the study region are carried out to gather the data on the accessibility of renewable energy sources, current energy consumption, etc. This study uses a questionnaire as a method for collecting specific knowledge from the field which consists of well-written questions and spaces given for the respondent to fill in with remarks or make checkmarks. This survey inquires more about the reality and opinions of participants in the study, through personal interactions, statements, and feedback. As seen below, the survey questionnaires are split into two sections, as current energy consumption and the personal information about the people. Survey about the current energy consumption includes existing preservation methods, fuels, transportation facilities, capacity and specification about the diesel Genset and information about the skill of the operators. The personal information questionnaire encompasses the daily household usage of various forms of energy, its potential and the utilisation of bioenergy resources. Each family's detailed information is gathered by personal interaction with the individuals. The accuracy of the data has been verified from the census data collected from government and non-government organizations, state and district administrations, veterinary dispensaries, etc.

1.2.1.3 Existing Collection and Preservation Methods for Milk

Among perishable agriculture products, milk is highly sensitive to time and temperature, and it needs immediate cooling to avoid spoilage in the supply chain. Moreover, small-scale dairy farming is popular in the study region. Therefore, this study is focused on this commodity for the analysis. The dairy industry sector primarily involves the dairy cooperative societies and corporate and government dairy farms which collect, refine, and sell milk and dairy products. To collect the milk from small-scale farming houses milkmen are employed, and the collected milk is processed by the milk cooperative societies situated in many spots. In a conventional method, it is sold directly to customers at the plant.

Normally farmers milk their livestock 2-3 times a day and they take the milk to nearby collecting centres where it is either processed or preserved.

Figure 1.2 Photographical view of dairy farms and agro preservation centre in the study region.

In this process, the milk produced is collected at the community level and immediately transported to the chilling centres or the bulk milk cooling units (BMC) available in selected places, where the milk is cooled to 4°C. The procurement of milk from farmers is conducted as per the design capacity of storage resources of the rural cooperative societies. Villages with BMC facilities have their milk obtained at 6- to 7-hour intervals twice a day, and it is chilled by the small-scale conventional cooling system. The capacity of the tank is 5,000 litres which can store the milk collected in one full-day storage or two milking cycles. One milking process yields between 2,000-3,500 litres of milk. Since the milk temperature in BMCs is kept at 4-6°C, the fermentation of milk is controlled, and the milk stays fresh and free from acidity and alcohol. The photographic view of a dairy farm and the milk chilling systems available in the study region is given in Figure 1.2.

1.2.1.4 Potential of Renewable Energy Sources

There are different types of biomass such as woody biomass, municipal waste and cow dung, which are available in the study region in a reasonable quantity. These types of biomass have the high potential to generate gaseous fuel either through thermochemical or the biochemical conversion process. Since the quality of gaseous fuels from these biowastes are different, they are grouped into three fuels such as producer gas, biogas and gobar gas which are obtained from biomass gasifier, municipal waste digester and cobra gas plant, respectively. This is also the reason for studying the potential for these wastes.

The main sources of Producer Gas Energy Sources (PGES) is tapioca stalk, Paddy straw, coconut shell, coconut rachises, wood chips, etc. The main BioGas Energy Sources (BGES) are Municipal Solid Waste, which is collected from domestic sectors. The amount of household waste produced from each family is 4 to 7 kg per day, and the quantity of waste needed to produce 1m^3 gas is expected as 24kg [29]. The major Gobar Gas Energy Sources (GGES) are identified from the available number of livestock population, and it is composed of the survey conducted from the rural farmers. It is stated that this waste is 13-16 kg/animal/day, 3.5-8.5 kg/animal/day and 0.2 kg/animal/day for buffalo, cattle, and goat, respectively, and the quantity of cow dung mandatory to produce 1m^3 gas is stated as 13kg [29, 30]. Calculation of the total dung produced annually is made through the production of animal dung per annum and the number of specific animals. Bioenergy is primarily used for heating and generating electrical energy in rural areas. Information in this regard has been collected from a minimum of ten villages in the study region, where farming is the primary source of capital. Table 1.1 shows some basic details of the study region.

Table 1.1 Details of the survey data.

Description of parameters	Values
The geographical area of the survey (sq.km)	23
Total population	4350
Number of households	1325
Livestock population	195
Land used for irrigation (%)	85
Producer gas energy resources (kg/day)	290-520
Biogas energy resources (kg/day)	1900-2250
Gobar gas energy resources (kg/day)	1120-1310
Volume of milk produced (l/day)	2900
Volume of agro products produced (kg/day)	4900

1.2.1.5 *Identification of Influential Parameters*

The research area has a dense population, and the main activity of the people is agricultural activities. Even though solar, wind and biomass are the major renewable energy sources, due to the climatic conditions and geographical nature, biomass is used for the agricultural-related energy needs. While there is an increasing need for resources to satisfy the daily requirements, the increasing constraint on land contributes to a reduction in the ability for energy supplies. In this case, either that the rural population has to settle for imported electricity, or perforce has to look for alternate sources to satisfy its basic needs. The volume of current resources in the rural area has a very important impact on energy use, demand and supply. Therefore, in terms of the scale of farmland, it is important to look very carefully at the activities of the population and demographics of the rural area.

Agriculture is the principal source of income in the field of study. Hence the scale of the operating resources is an important parameter for evaluating the market and power production. Moreover, the energy sources from humans, animals, farmyard, etc., are to be taken into consideration in the renewable energy generation. Although the worth of the farm assets is not directly related to energy consumption, data are collected about the nature and extent of the investment to determine the households' economic status. The worth of a farm's properties has been calculated as investing in land, cattle, farm equipment and farm outfits.

Cattle commodities are among the most significant aspects in an agricultural society, which consumes energy and produces energy sources besides agricultural outputs. The amount of energy consumption and production vary according to the nature of livestock. The livestock population is categorized into various groups for the assessment of energy consumption and production. The intensity of cropping is another critical factor deciding the energy required to carry out agriculture activities. Depending on the infrastructure facilities, the field crops are grown in one, two or three seasons in the study area. The energy demand and supply would increase accordingly for a farm that cultivates field crops in more than one season. These factors are considered for the calculation of biomass potential.

1.2.1.6 *Load/Demand Assessment*

Based on milk production, the number of livestock and potential of renewable energy resources, the cooling load to preserve the milk and other dairy products in the vapour absorption system and the necessary thermal energy to run the chilling equipment shall be determined.

The volume of dairy products such as milk provided in remote regions is estimated by the subsequent equation:

$$\textit{Volume of milk} = (\textit{No. of livestocks} \times \textit{milk produced/livestock}) \quad (1.1)$$

Generally, each cow produces 7-18 litres of milk per day [31]. The required mass of dairy products is estimated by

$$\textit{Mass of milk} = \textit{Volume of milk} \times \textit{Density of milk} \quad (1.2)$$

Based on the literature, density of milk is taken as 1027 kg/m^3 [29, 30]. The cooling load (Q_e) from the evaporator of the vapor absorption cooling system is obtained by

$$\textit{Cooling load } (Q_e) = \textit{Mass of milk} \times C_{p,m} \times (T_{milk} - T_{evaporator}) \quad (1.3)$$

where, $C_{p,m}$ is the specific heat of milk and it is taken as 3.93 kJ/kgK [29, 30].

Based on the cooling load demand and the need expected heat input to the generator of the refrigeration system, the required quantity of producer gas, biogas, and gobar gas sources are determined. The cooling capacity of the evaporator is calculated from equation 1.3. The heat load of the generator to run the absorption system is estimated on the basis of the normal COP of the absorption chiller and the necessary amount of cooling effect of the absorption cooling system by using the following relation:

$$Q_{gen} = {}^{Qe}/COP_{chiller} \quad (1.4)$$

The amount of renewable energy sources available shall be determined by using the relation below:

$$m_{PGES} = \frac{Q_{gen}}{HV_{PGES} \times \eta_{c,PGES}} \quad (1.5)$$

$$m_{BGES} = \frac{Q_{gen}}{HV_{BGES} \times \eta_{c,BGES}} \quad (1.6)$$

$$m_{GGES} = \frac{Q_{gen}}{HV_{GGES} \times \eta_{c,GGES}} \quad (1.7)$$

where, m_{PGES}, m_{BGES} and m_{GGES} are defined as the mass of PGES, BGES and GGES, respectively. Similarly HV_{PGES}, HV_{BGES} and HV_{GGES} are defined as the heating values and $\eta_{c,PGES}$, $\eta_{c,BGES}$ and $\eta_{c,GGES}$ are the conversion efficiency of the respective energy source.

The amount of each available renewable energy source and the quantity needed to supply the required heat to the cooling system in the study region is determined based on the survey data. It is observed that the quantity of renewable energy sources is less than the energy needed to meet the cooling requirement; therefore all three types of gaseous biofuels are needed to operate the chilling system for the preservation of agro products in the study region. However, the appropriate mix of these sources is identified to run the system without the starvation of heat energy.

In this study, different mixtures PGES, BGES and GGES, are studied, besides the possibility of utilizing them to operate a chilling system in the study region. To study the impact of the variation in the availability of energy sources, two terms, namely base element and mixture element, are used. The base element is any single source of energy, and the mixture element is a mixture of two different sources of energy. The constant energy ratio is maintained in the mixture element as shown in Table 1.2. The proportion of the mixing element is 25%, 50% and 75%. In all energy combinations, the proportions of the renewable energy sources in the mixture element are kept constant for a specific mix of energy sources.

Table 1.2 Ratios of renewable energy sources in the selected mixtures.

Sl. no.	Renewable energy sources in the mix	Mixing ratio
1	PGES: BGES+GGES	1: 0.5+0.5
2	PGES: BGES+GGES	1: 0.75+0.25
3	PGES: BGES+GGES	1: 0.25+0.75
4	BGES: PGES+GGES	1: 0.5+0.5
5	BGES: PGES+GGES	1: 0.75+0.25
6	BGES: PGES+GGES	1: 0.25+0.75
7	GGES: PGES+BGES	1: 0.5+0.5
8	GGES: PGES+BGES	1: 0.75+0.25
9	GGES: PGES+BGES	1: 0.25+0.75

1.2.2 Modelling and Simulation of a Hybrid Renewable Energy–Based Cooling System

The performance and economic advantages of the proposed hybrid-energy-operated vapour absorption cooling system has been studied with a suitable mathematical model, which is developed in MATLAB. The factors which have a significant impact on the decision making are incorporated in the analysis. Since the minimum expected capacity of the cooling system for the regions is estimated as 35 kW, the same has been used to derive the performance characteristic and economic factors. The assumptions taken for this study are: steady-state conditions; constant heating values of renewable energy sources; constant conversion performance of specific renewable energy source; constant performance of absorption chiller; constant cooling demand; labour cost and fuel price variation during the period of calculation is negligible. Suitable steps are also used to calculate the uncertainty and identify the most influential parameters for the implementation of this methodology for the preservation needs in the selected region.

1.2.2.1 System Description

The entire chilling system with renewable energy sources has been divided into two subsystems as shown in Figure 1.3.

Subsystem 1 (Renewable energy conversion system): The energy available in the different types of biomass, considered as feedstock material, is converted to thermal energy for heating the generator of the absorption chiller. For the energy conversion cycle, the specific conversion devices such as biomass gasifier, biogas digester and gobar gas digester are used. The selection of these devices is decided from the availability of energy sources and the expected contribution of these energy sources in the hybrid energy, besides the cooling load demand.

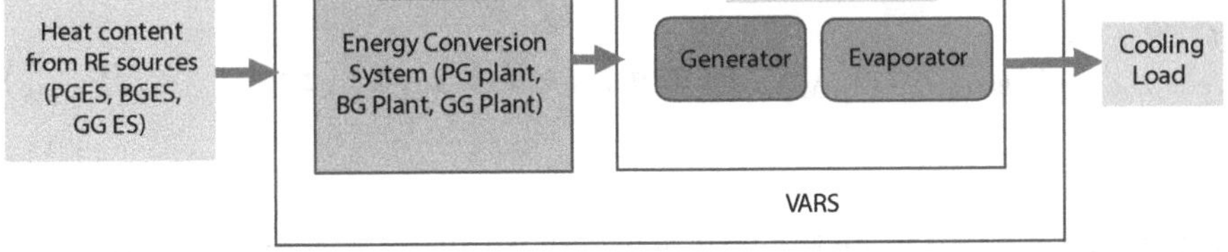

Figure 1.3 The proposed hybrid renewable energy–operated refrigeration system.

Subsystem 2 (Vapour absorption chilling system): In this study, the vapour absorption chiller working with the lithium bromide-water pair has been identified as per the recommendation given in the literature. The evaporator of the absorption chiller is to deliver the required amount of cooling load for the preservation of agro products. In this study, the milk produced in the selected region has been used to calculate the evaporator capacity.

1.2.2.2 Energy Modelling

The required quantity of heating and cooling energy is evaluated based on Equations 1.3 and 1.4. Thermal energy delivered to the generator is evaluated from the heat energy produced from the proper mixing of available renewable energy sources, which is converted into thermal energy with appropriate energy conversion devices such as biomass gasifiers and digesters. The thermal energy calculations are performed from the conversion efficiencies of the devices involved in the renewable energy conversion systems. The thermal energy delivered from the energy conversion systems is calculated based on the following equations.

$$Q_{PGES} = m_{PGES} \times HV_{PGES} \times \eta_{c,\,PGES} \tag{1.8}$$

$$Q_{BGES} = m_{BGES} \times HV_{BGES} \times \eta_{c,BGES} \tag{1.9}$$

$$Q_{GGES} = m_{GGES} \times HV_{GGES} \times \eta_{c,GGES} \tag{1.10}$$

where, Q_{BMES}, Q_{BGES} and Q_{GGES} are defined as the heat from PGES, BGES and GGES respectively. Similarly m_{BMES}, m_{BGES} and m_{GGES} are defined as the mass, HV_{BMES}, HV_{BGES} and HV_{GGES} are referred to as the heating value and $\eta_{c,BMES}$, $\eta_{c,BGES}$ and $\eta_{c,GGES}$ are the conversion efficiencies of the respective energy sources.

The overall efficiency of the system of the proposed hybrid renewable energy system is evaluated from the following equation.

$$OES = \frac{Q_e}{(x.m_{BMES} \times HV_{BMES}) + (y.m_{BGES} \times HV_{BGES}) + (z.m_{GGES} \times HV_{GGES})} \tag{1.11}$$

where x, y, and z denote the proportion of producer gas, biogas, and gobar gas energy sources in a hybrid renewable energy system.

1.2.2.3 *Economic Modelling*

An optimum mixture of hybrid renewable energy-operated absorption cooling systems can be found as the best way to satisfy the cooling energy requirement with maximum OES and the highest economic feasibility. The economic parameters considered in the study are Capital Investment Cist (CIC), Operating Cost (OC), Simple Payback Time (SBPT), Net Present Cost (NPC), Total Annual Cost (TAC), Annualised Capital Cost (ACC) and Life Cycle Cost (LCC). The procedure for calculation of the important economic parameters for this study is described in Table 1.3 [30, 32].

A software algorithm is established in the MATLAB software, and the optimal design methodology is selected to analyse the overall efficiency of the system and economic factors of the various mixture compositions of energy sources. The input values to the parameters selected for this study are selected from the survey and literature. The cost components, criteria to use various input parameters and the efficiency of the chilling system are also selected with appropriated validation. The procedures followed and the various stages of the modelling and simulation study are represented in the flow chart depicted in Figure 1.4.

The design process involves preliminary data to predict a suitable hybrid design that meets local conditions and needs. The input is used to configure the simulation that represents the characteristic details for the region at which a hybrid system is proposed for implementation. The information obtained regarding the demand for cooling loads, accessible renewable energy sources, characteristics of these renewable energy sources, environmental factors and product specifications, including operational characteristics and costs, are used. Availability may vary between the environmental conditions and the nature of the regions. Demand management is also an important factor to be considered in the development of demand analysis because it will dramatically minimize the device cost if the demand is optimised to the available energy. The analytical model can predict OES, CIC, OC, SPBT, LCC and NPV of multiple mixtures of energy sources in the region in this study.

1.2.2.4 *Simulation and Performance Evaluation*

In the simulation model developed in MATLAB, the input parameters are given with appropriate values which are either derived from the field-survey or standard values available in the literature. The proposed combinations of the three types of biomass energy sources in hybrid renewable energy are also defined. The main renewable energy sources PGES, BGES

Table 1.3 Procedure for analysing the economic parameters used in this study.

Economic parameters	Calculation procedure
Capital Investment cost (CIC)	CIC of producer gas plant + CIC of biogas plant + CIC of gobar gas plant + CIC of biogas burner + CIC of gobar gas burner + CIC of biogas and gobar gas heat exchanger – government subsidies
Operating Cost (OC)	The costs involved in association with the energy source, operation, maintenance, energy conversion systems, cooling system, Labour and Depreciation. (maintenance cost, lifetime and the annual interest are taken as 2% of the CIC, 20 years, and 12% respectively. The costs of PGES, GGES and BGES are taken as INR 1.40/kg, INR 0.60/kg and INR 0.80/kg)
Simple payback Time (SPBT)	$\frac{Incremental\ value\ of\ CIC}{Annual\ profit}$
CIC of diesel-powered compression chiller	Cost of diesel Genset + Cost of compression chiller
OC of diesel-powered compression chiller	Cost of diesel + maintenance cost of compression chiller + labour cost to handle the system + depreciation value
Annualised Capital Cost (ACC)	Capital Investment cost (CIC) × Capital Recovery Factor (CRF)
Capital Recovery Factor (CRF)	$CRF = \frac{i.\,(i+i)^n}{(1+i)^n - 1}$ where i is the annual interest rate, n is the lifetime of the component.
Total Annual Cost (TAC)	Annualised Capital Cost (ACC) + Annual operation and maintenance cost
Net Present Cost (NPC)	$NPC = \left[S.\left(\frac{(1+i)^n - 1}{i(1+i)^n} \right) \right] - CIC$ where S as benefits at the end of the project

(*Continued*)

Table 1.3 Procedure for analysing the economic parameters used in this study. (*Continued*)

Economic parameters	Calculation procedure
Life Cycle Cost (LCC)	$LCC = CIC + PWV_{OC} + PWV_{ReC} + PWV_{SV}$
Present Worth Value of operating cost (PWV_{OC})	$PWV_{OC} = OC.\left[\frac{(1+i)^n - 1}{i(1+i)^n}\right]$
Present Worth Value of replacement cost (PWV_{ReC})	$PWV_{ReC} = ReC.\left[\frac{(1+i)^n - 1}{i(1+i)^n}\right]$
Present Worth Value of salvage value (PWV_{SV})	$PWV_{SV} = SV.\left[\frac{(1+i)^n - 1}{i(1+i)^n}\right]$ Here the salvage value and replacement cost of the system after 20 years is estimated by assuming 7% of CIC and 1% of CIC of the system, respectively.

and GGES are used in this approach to satisfy the required heat to meet its demand for the operation of the cooling system. The process parameters included in the simulation study are the availability of renewable energy sources in the study area, the physical characteristics of available renewable energy sources, meteorological data, process parameters of VARS and the costs components that are either provided from the suppliers or collected from the field survey. The important input parameters considered in the study are volume of the product to be preserved, standard COP of the absorption chiller, heating value of the PGES, BGES and GGES, conversion efficiency of the PG plant, BG plant and GG plant, CIC of the energy conversion system, cost of the identified renewable energy sources per kg, government subsidy, cost of the labour, depreciation value, etc.

The major function of the simulation is to identify the proper hybrid system to meet local needs and requirements. The data has been used to carry out a simulation that describes the specific data for the region where a hybrid system is being considered. Information gathered on the demand for cooling load, available renewable energy sources, the properties of renewable energy sources, climate conditions and definitions of the identified components, including the modes of operation and costs shall be used for this purpose. Economic parameters such as capital investment

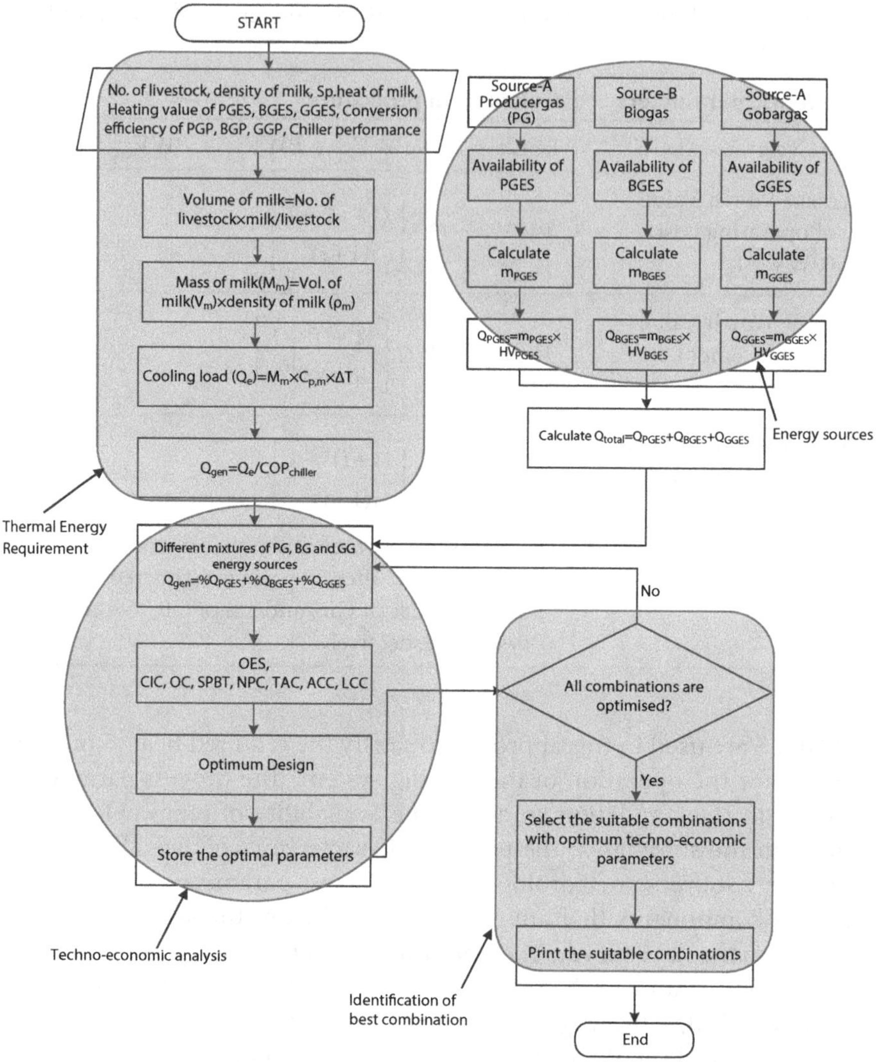

Figure 1.4 Analysis procedure for the proposed system.

cost, operating cost, simple payback time, net present cost and life cycle cost, etc., are also considered in this study. Demand may vary between the climate patterns and the nature of the regions selected. Demand management is another important factor to consider the expected cooling load as it can significantly reduce the cost involved in the system if the demand is tailored to the available energy. As stated in the flow chart, the simulation steps are followed in two sections and the combined outcomes are used to further calculate the economic parameter, techno-economic decision-making results, overall system performance for the implementation of the selected mixtures, etc. The convergence of the simulation is defined based on the realistic values of the techno-economic parameters.

1.3 Results and Discussion

The effects of techno-economic modelling and simulation of a hybrid renewable energy-powered chilling system in the study region are presented. The average milk yield in many of the areas surveyed is about 5,000 L per day that can be cooled in 4-6 hours to the preservation temperature at 4-5°C. Hence in this analysis, a refrigeration system with 10TR cooling capacity was considered. The output performance parameters as the overall efficiency of the system and specified economic factors are analysed in this chapter.

In the study region, the average quantity of producer gas, biogas and gobar gas energy sources are identified as 290-520 kg/day, 1900-2250 kg/day and 1120-1310 kg/day respectively while the required cooling load is 28kW. The average cooling load of a hybrid energy–operated cooling system is measured as 12.4kW, 16.6kW and 14.7kW, respectively, using the maximum utilisation of available energy sources in different mixtures as PGES-BGES, PGES-GGES and BGES-GGES. Therefore, to meet the cooling load, a minimum of three identified renewable energy sources must be combined.

1.3.1 Overall Efficiency of the System

The overall efficiency of the system for the different renewable energy combinations, suitable in the study region, is given in Table 1.4. The base element is used as any single source of energy, and the mixture element is a mixture of two different sources of energy. The proportion of the mixture element mixed from 25%, 50% and 75%, respectively. The term EMR (Energy Mixing Ratio) is used in all the illustrations to denote the various proportions between the base and mixture components. In the figures, the percentage of base energy component in the EMR increases from left to right, while the mixture energy component increases from right to left in the X-axis.

The variations of OES with the variation of EMR for all the mixtures are shown in Figures 1.5-1.7. The constant energy ratio in the mixture element is also shown in Table 1.2. In all energy combinations, the proportions of the renewable energy sources in the mixture element are kept constant for

Table 1.4 Mixtures of renewable energy sources with various energy proportions.

Combination	Base energy source	Mixture energy sources	The ratio of mixture components	OES
MR1	PGES	BGES:GGES	0.50:0.50	0.16-0.24
MR2	PGES	BGES:GGES	0.75:0.25	0.14-0.24
MR3	PGES	BGES:GGES	0.25:0.75	0.17-0.24
MR4	BGES	PGES:GGES	0.50:0.50	0.14-0.21
MR5	BGES	PGES:GGES	0.75:0.25	0.14-0.23
MR6	BGES	PGES:GGES	0.25:0.75	0.14-0.20
MR7	GGES	PGES:BGES	0.50:0.50	0.17-0.19
MR8	GGES	PGES:BGES	0.75:0.25	0.19-0.20
MR9	GGES	PGES:BGES	0.25:0.75	0.15-0.19

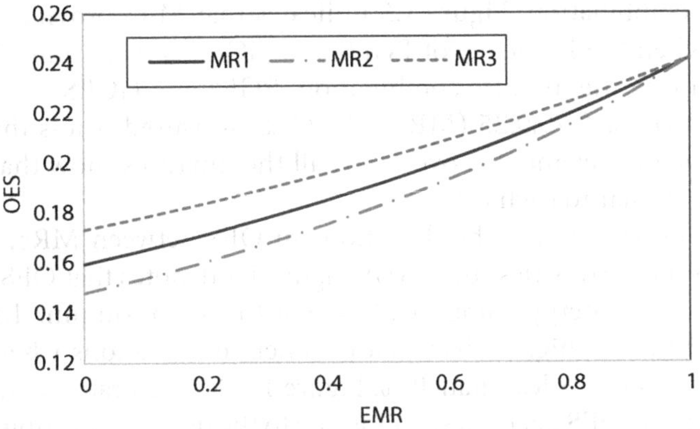

Figure 1.5 Overall efficiency of the system with the EMR of MR1, MR2 and MR3.

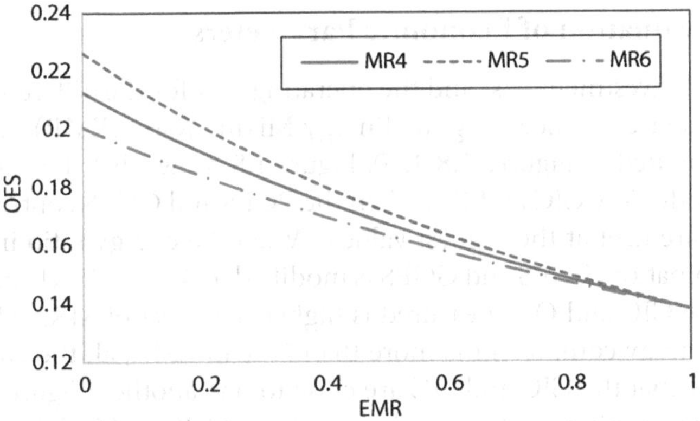

Figure 1.6 Overall efficiency of the system with the EMR of MR4, MR5 and MR6.

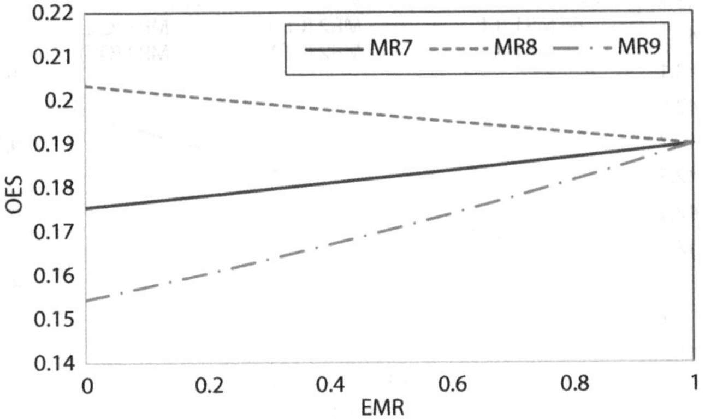

Figure 1.7 Overall efficiency of the system with the EMR of MR7, MR8 and MR9.

a specific combination. Figure 1.5 indicates that MR3 provides the highest OES when the influence of PGES and GGES is maximum. When the energy ratio in the mixture combination, BGES and GGES is modified to 0.5:0.5 (MR1) or 0.75:0.25 (MR2), the OES measured is less than that of MR3. If the base element is over 80%, all the mixtures show that the OES values are similar to each other.

Figure 1.6 shows that the deviations in OES between MR4, MR5 and MR6 are very narrow (less than 8%). Figure 1.7 denotes that OES improves when the basic energy component is kept to the maximum. The OES of MR8 is 20 to 35% higher than that of other combinations when the base energy component is less than 45%. Figure 1.5-1.7 illustrates that, when the BGES rises, the OES decreases. This is due to the deprived heating value and the conversion efficiency of BGES as compared with the PGES and GGES.

1.3.2 Evaluation of Economic Parameters

The capital investment cost and the operating cost for the different mixtures of energy sources concerning the Energy Mixing Ratio (EMR) in the study region is plotted in Figures 1.8-1.10. Figure 1.8 shows that the combination MR3 provides low CIC and OC when the PGES and GGES contributions in the EMR are kept at the highest values. When the energy ratio in the mixture combination, BGES and GGES is modified to 0.5:0.5 (MR1) or 0.75:0.25 (MR2), the CIC and OC measured is higher than that of MR3. However, if the base energy component is more than 75% in EMR, all the energy mixtures show that the CIC and OC are close to one another. Figure 1.9 shows that the energy mixture MR6 provides the lowest CIC and OC than MR4 and

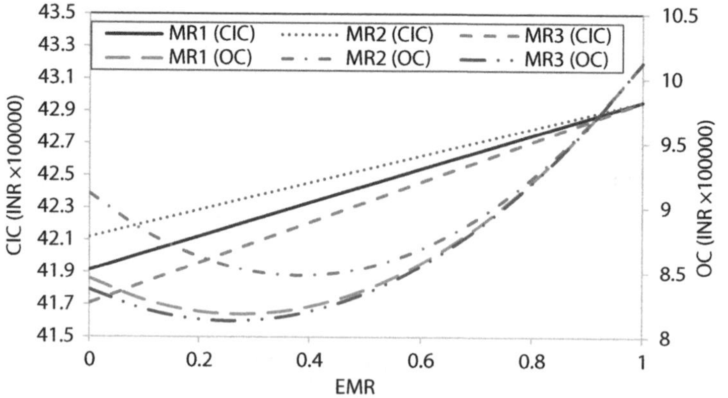

Figure 1.8 Variations in CIC and OC with the EMR of MR1, MR2 and MR3.

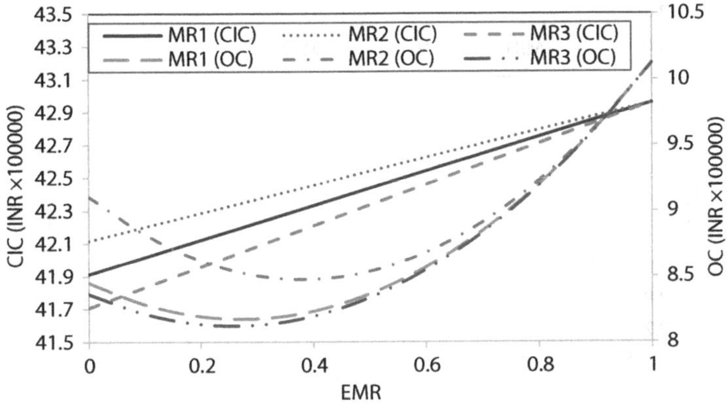

Figure 1.9 Variations in CIC and OC with the EMR of MR4, MR5 and MR6.

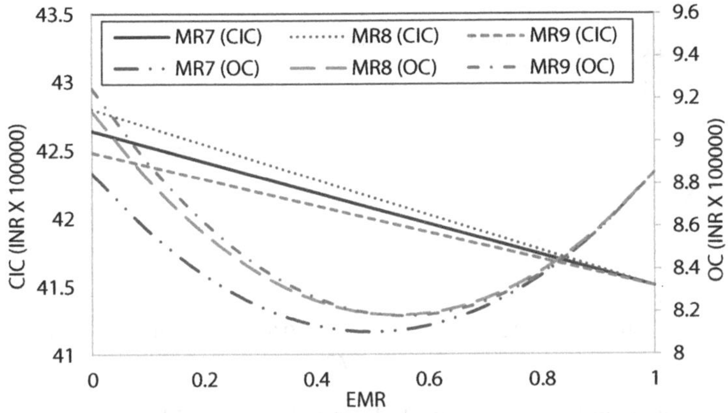

Figure 1.10 Variations in CIC and OC with the EMR of MR7, MR8 and MR9.

MR5, when the BGES and GGES are maintained at the maximum. When the GGES performs as a base energy component and the PGES and BGES contributions are maintained at 25% and 75%, respectively, the CIC of MR9 is 12% lower when compared to MR7 (Figure 1.10). However, the OC of MR9 is 16% higher than that of MR7. It is also seen that the CIC of all three energy combinations are close to each other when the ratio of base energy component is raised above 70%. A similar trend is also observed in the OC. The results indicate that the CIC and OC have low values when the GGES is selected as a base energy component, which is also indicated in MR7-MR9.

Figures 1.11-1.13 show that the energy mixtures MR7-MR9 have the least SPBT as compared to MR1-MR6. When the GGES as a base energy element, the lowest SPBT has been observed. Therefore, GGES can be

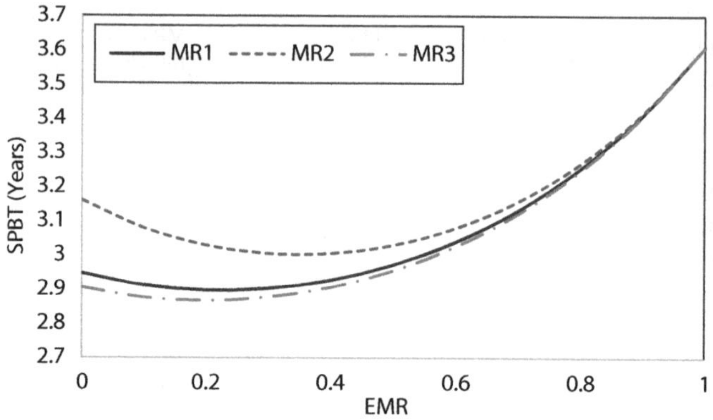

Figure 1.11 Variations in SPBT with the EMR of MR1, MR2 and MR3.

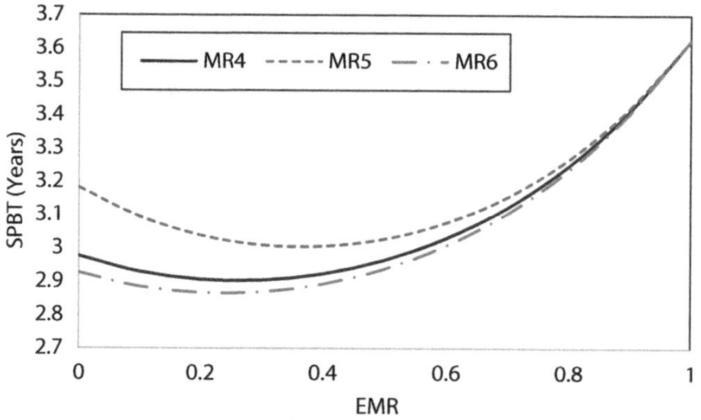

Figure 1.12 Variations in SPBT with the EMR of MR4, MR5 and MR6.

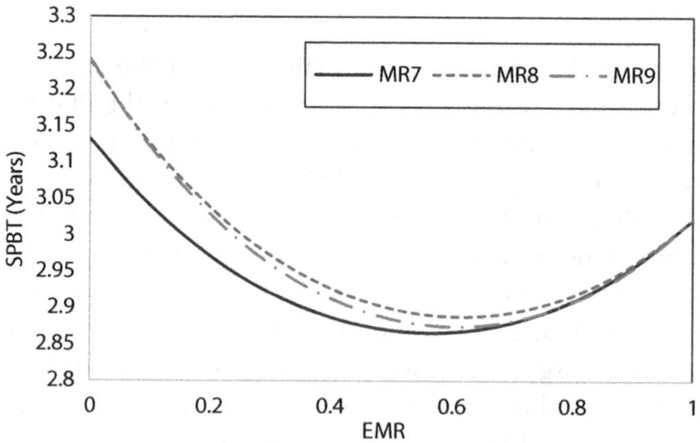

Figure 1.13 Variations in SPBT with the EMR of MR7, MR8 and MR9.

nominated as the base energy element to attain the least value of SPBT. When compared to the SPBT, MR7 shows the lowest value; this is because of the equal influence of PGES and BGES in the mixing proportion. When the value of the base energy component rises above 75%, the SPBT of all the energy contributions is similar to each other.

Figure 1.14-1.16 shows that the net present costs of MR1-MR9 are positive irrespective of EMR. Moreover, the highest NPC is seen, when the EMR is up to 75%, and it is close to one another when the EMR is above

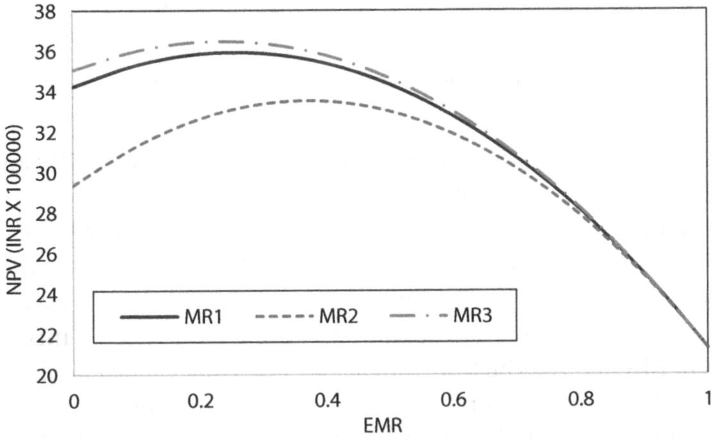

Figure 1.14 Variations in NPV with the EMR of MR1, MR2 and MR3.

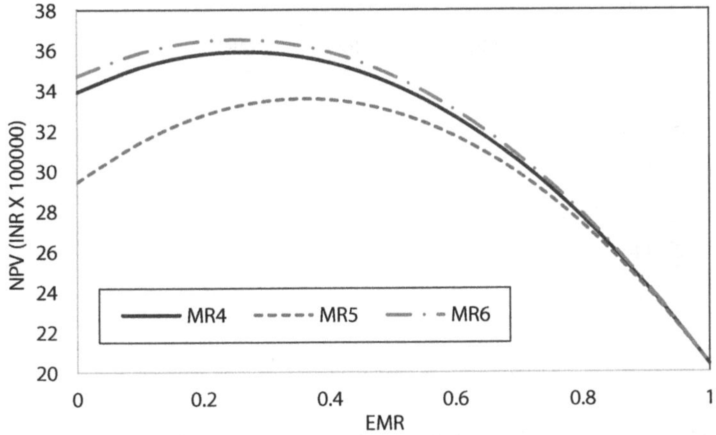

Figure 1.15 Variations in NPV with the EMR of MR4, MR5 and MR6.

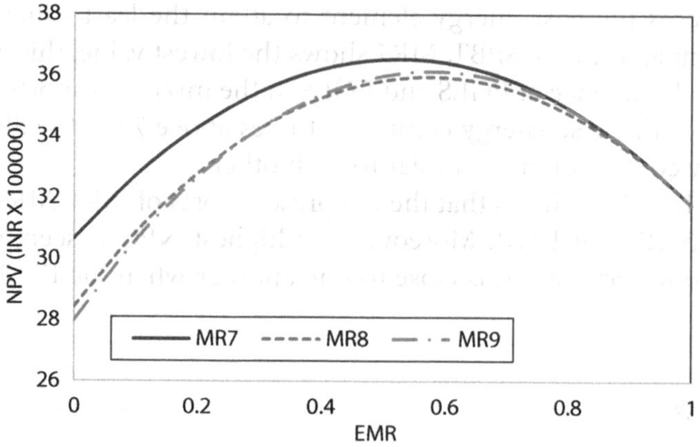

Figure 1.16 Variations in NPV with the EMR of MR7, MR8 and MR9.

75%. Figure 1.16 shows that the NPC of MR7 is higher than MR8 and MR9 which is due to the impact of GGES.

The life cycle cost of various energy mixtures, MR1-MR9 is shown in Figures 1.17-1.19. The energy combinations MR7, MR8 and MR9 have the lowest LCC as shown in Figure 1.19; this is due to the involvement of GGES as a base energy element. When the LCC of all the mixtures are compared, MR7 shows the lowest LCC, which is due to the equal contribution of PGES and BGES in the energy mixing proportion.

The annualised capital cost and total annual cost are evaluated for the different energy combinations and plotted in Figures 1.20-1.22. The ACC

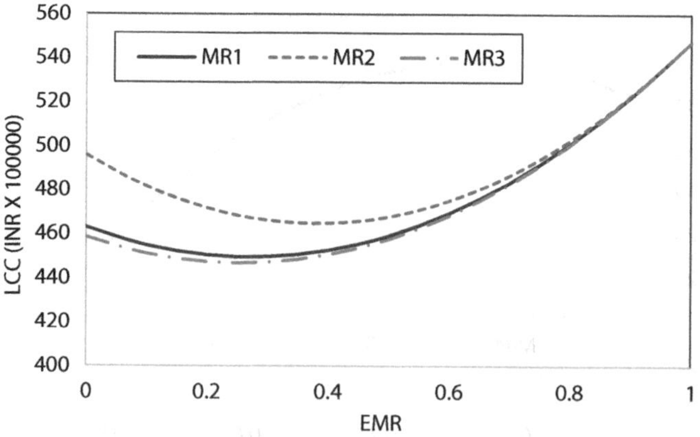

Figure 1.17 Variations in LCC with the EMR of MR1, MR2 and MR3.

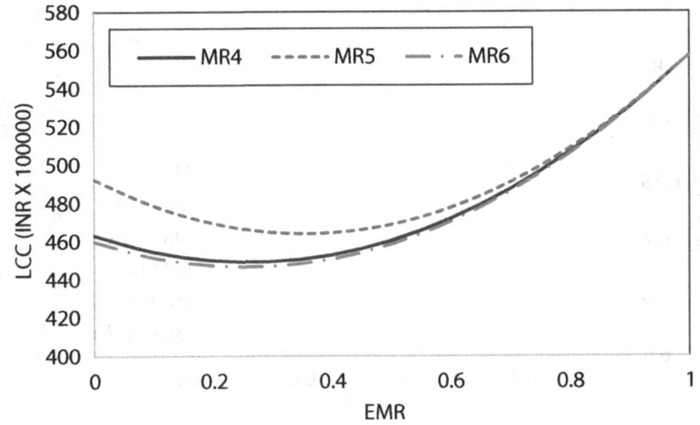

Figure 1.18 Variations in LCC with the EMR of MR4, MR5 and MR6.

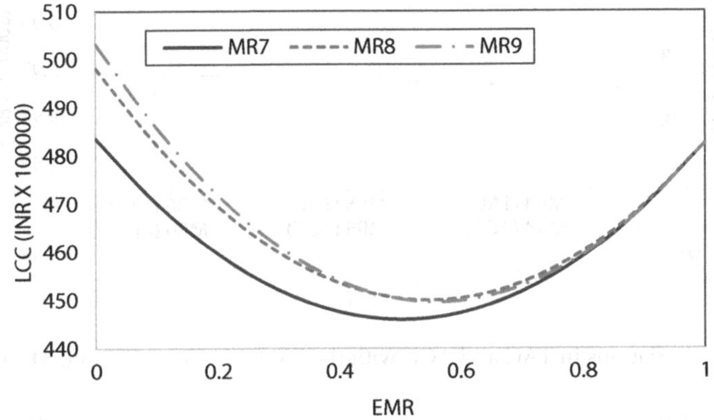

Figure 1.19 Variations in LCC with the EMR of MR7, MR8 and MR9.

varies from INR 58×10^4-60×10^4. Since ACC indicates the average expense of running a device over its entire life cycle, it should be reduced and at the same time, marginal profit for the mixture should be achievable. MR7 appears to be the most fitting formulation to meet these conditions. TAC is used to forecast and estimate the annualized expense of each portion of the system and the expense of each conversion device, and hence it should be held to a minimum. Figures 1.20-1.22 show that the TAC for all the energy mixtures varies from INR 67×10^4- 69×10^4. The lowest TAC is observed in the combination of MR7-MR9, which is similar to ACC.

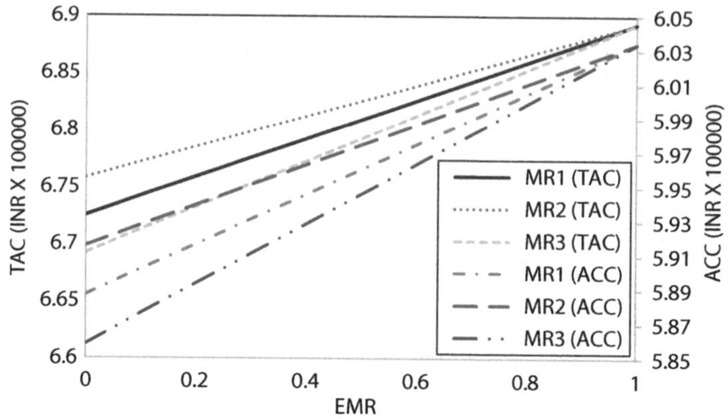

Figure 1.20 Variations in TAC and ACC with the EMR of MR1, MR2 and MR3.

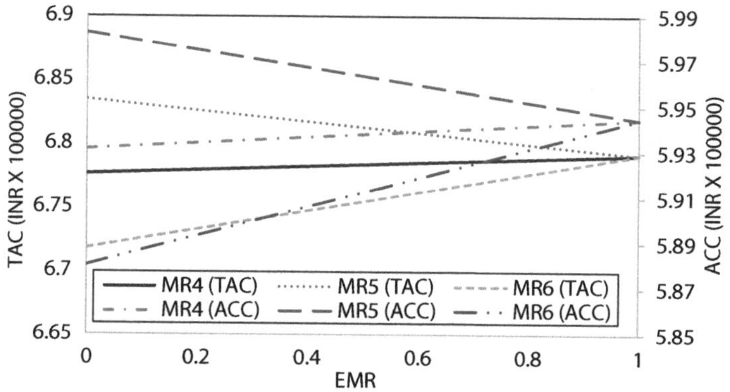

Figure 1.21 Variations in TAC and ACC with the EMR of MR4, MR5 and MR6.

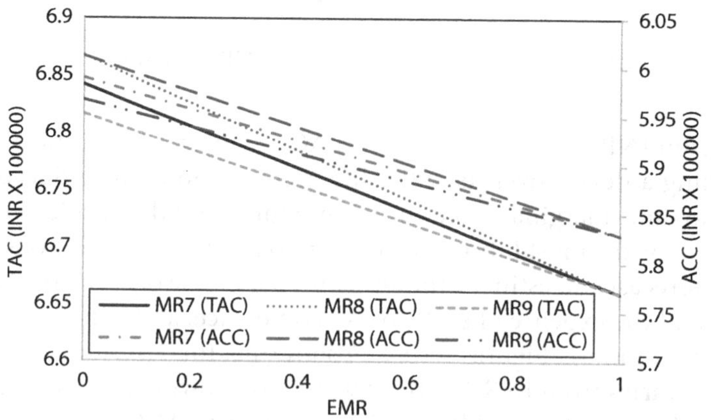

Figure 1.22 Variations in TAC and ACC with the EMR of MR7, MR8 and MR9.

1.3.3 Techno-Economic Study

The techno-economic analysis aims to find the relation between the overall efficiency of the system and the different financial variables within the defined system that will give an optimum overall efficiency with financial benefits. The possible energy combinations identified in the study region along with their OES and the economic parameters are studied in this work. The observations show that the energy mixture MR3 provides the maximum OES when PGES and GGES work with the maximum efficiency; however, a high CIC, OC, SPBT and LCC are found for MR3. The lowest value of OES and the highest value of CIC, OC and the SPBT are also found in other mixtures MR4 and MR6. Hence, these energy combinations are not feasible in the study region.

The energy mixture of MR7 to MR9 shows reasonable OES and financial benefits. The OES and other financial advantages of MR9 are less than that of MR8. When MR7 and MR8 are compared, the financial benefits of MR7 are 8-12% less than MR8. Among the three mixtures, MR8 has a minimum of 18% higher OES, and hence it has been selected as a suitable energy combination for the identified study region. Also, the NPC value of the energy combination MR8 is positive for all possible EMR, so MR8 has been identified as an appropriate renewable energy combination to implement completely renewable energy-operated vapour absorption food preservation system in the study region.

1.3.4 Sensitivity Analysis

The sensitivity analysis is the authentic tool to predict the fluctuation in the results due to error in various observations and data collection; besides, the error in the modelling methods can be identified and the accuracy of the predictions would be confirmed. In the simulation of any system, only some parameters have a significant influence on the predicted results and the same can be traced from this analysis. In this study, sensitivity analysis is used to determine the effect of variations in input variables such as cooling demand, the quantity of energy, fuel price, the performance of various systems in the conversion process, rate of interest, etc., on the calculated results such as OES, CIC, OC and SPBT.

Figure 1.23 shows that the variations in cooling demand and the efficiencies of energy conversion plants have a significant influence on the OES. The efficiencies of the three energy conversion devices also have a serious impact on the OES. The maximum impact is observed in producer gas plant followed by biogas digester and gobar gas digester. This trend

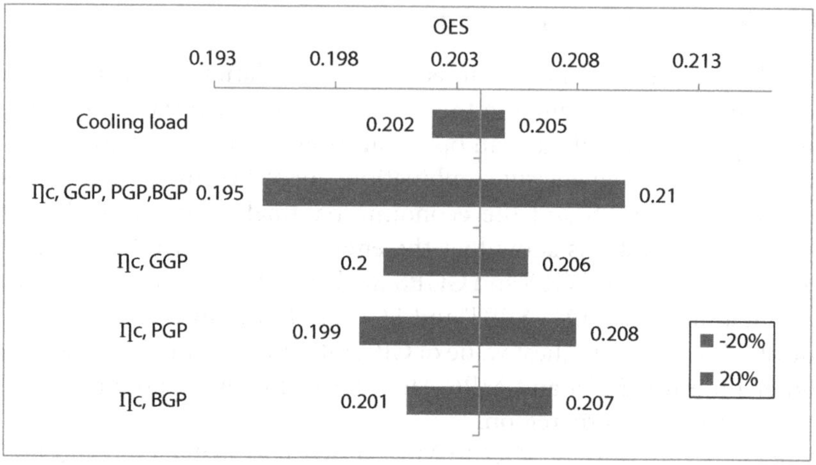

Figure 1.23 Sensitivity analysis of overall efficiency of the system.

proves that even with a small improvement or fall in the conversion efficiencies, a significant change in the overall efficiency of the system can be possible.

Figure 1.24 depicts the impact of various parameters on the variations in capital investment cost of the system. It is observed that the capital cost of absorption chiller has a major impact on the CIC followed by the CIC of GGP. A 20% variation in the CIC of VARS can influence the CIC of the system by 2.5 to 3.5%, whereas the variation in the CIC of BGP has a negligible impact. In general, the capital cost variation of the producer gas

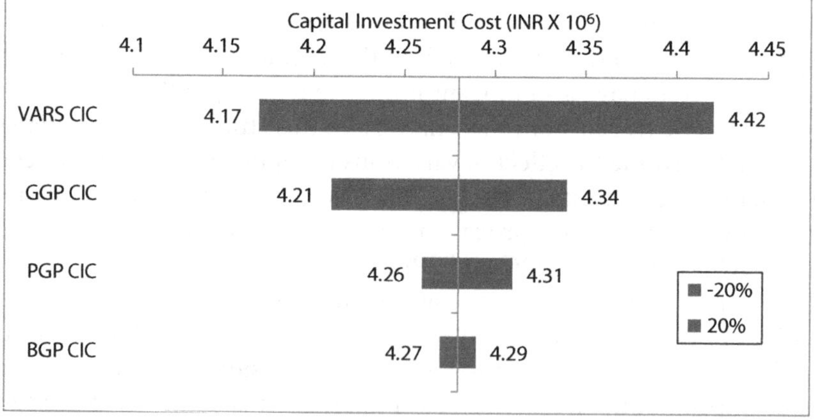

Figure 1.24 Sensitivity analysis of capital investment cost of the system.

plant, biogas plant and gobar gas plant have a low incidence in the deviation range.

Figures 1.25 and 1.26 depict the impact of the variation of influential parameters on the operating cost and the simple payback time of the proposed hybrid energy cooling system. The variation of the CIC of energy conversion plants and costs of energy sources have a serious influence on the OC and SPBT. It is found that the interest rate variance of ±20% has a substantial effect on the OC and SPBT of the system. An increase in the

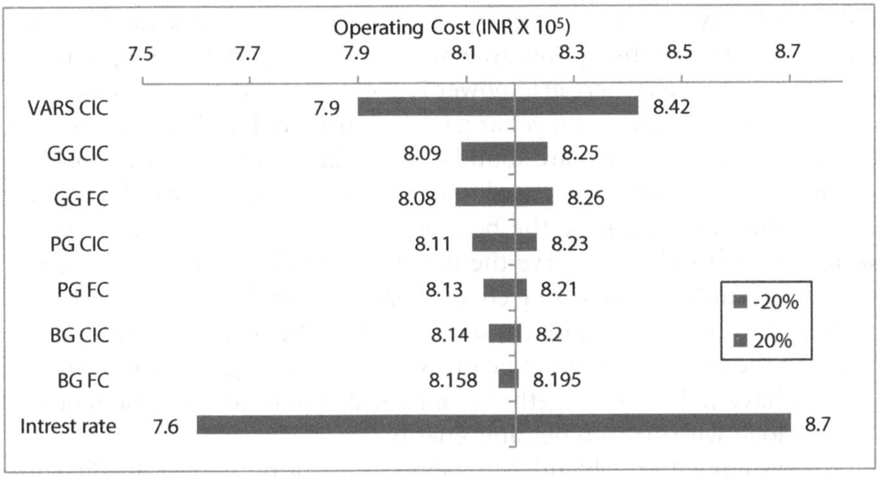

Figure 1.25 Sensitivity analysis of the operating cost of the system.

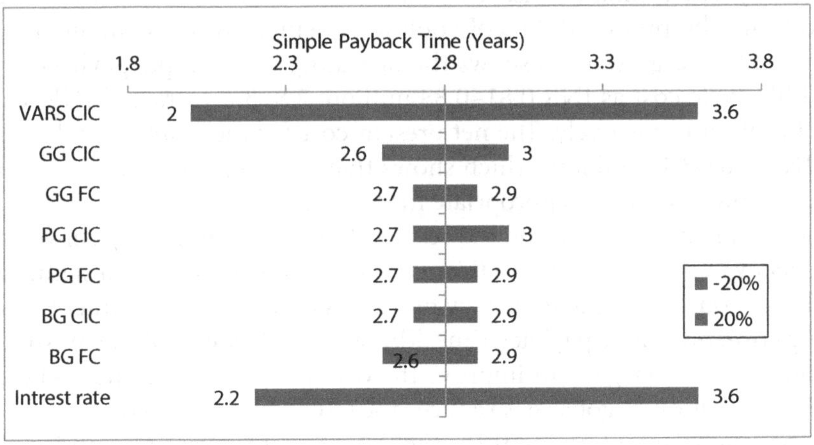

Figure 1.26 Sensitivity analysis of simple payback time of the system.

capital cost of gobar gas plant, producer gas plant, biogas plant and energy supply prices is responsible for a small occurrence in the variability spectrum of OC and SPBT (less than 1.5%).

1.4 Conclusions

The proposed work focuses on developing an environmentally sustainable, 100% energy-efficient cooling system for the chilled-preservation agricultural foodstuff in agro-tropical areas. The analysis has been carried out for a different mix of renewable energy sources to supply heat for the operation of a vapour absorption system to preserve perishable agricultural products in regions where grid power is not feasible. The gaseous fuel such as producer gas, biogas and gobar gas are considered in this analysis since the sources for these fuels are abundantly available in the study region. The techno-economic analysis is used to find an appropriated mix of the above energy sources to manage the heat energy needs of the cooling system which is preferred to preserve the perishable foodstuff at 4 to 5°C. From the overall study, the following conclusions are also derived.

Since the cooling load demand expected in the regions cannot be satisfied by the heat from any single or two types of energy, all three types of sources have to be used together as a hybrid system to meet the required cooling load with 100% renewable energy.

The system with the blend, MR8 shows the optimum overall efficiency, capital investment cost, operating cost, simple payback time and life cycle cost. Therefore, this energy mixture can be preferable over the other proposed hybrid combinations. It offers the overall efficiency between 0.19 and 0.24. The proposed 35 kW cooling system with this combination of renewable energy sources shows the operating cost, simple payback time and life cycle cost as INR 0.81–0.88 million, 2.8–3.2 years and INR 44.5–48.4 million respectively. The net present cost for the system is calculated as INR 2.84-3.12 million, which shows that the proposed mixture of bioenergy sources is more appropriate for the study region.

The sensitivity analysis on the results shows that the energy conversion efficiency and interest rate are the major parameters that have a significant influence on the overall efficiency of the system, capital investment, operating cost and simple payback time. The study indicates that the maximum usage of producer gas can improve the overall system efficiency, whereas the contribution of gobar gas reduces the CIC, OC and SPBT.

The major intention of this study is to raise awareness and confidence among the various stockholders of tropical regions to establish a completely renewable energy–operated hybrid system for local application. Moreover, this study can pave the way for researchers and rural entrepreneurs to work further on implementing this methodology for various applications in remote regions with economic merits; besides, it provides some useful information on the optimal planning and utilization of hybrid renewable energy systems in the preservation of perishable agricultural produce.

References

1. Edwin, M., Joseph Sekhar, S., Thermal performance of milk chilling units in remote villages working with the combination of biomass, biogas and solar energies. *Energy*, 91, 842-851, 2015.
2. Mala Ramesh, Rajeshwer Prasad Saini, Dispatch strategies based performance analysis of a hybrid renewable energy system for a remote rural area in India. *Journal of Cleaner Production*, 259, 120697, 2020.
3. Jiah Yu, Jun-Hyung Ryub, In-beum Lee, A stochastic optimization approach to the design and operation planning of a hybrid renewable energy system. *Applied Energy*, 247, 212-220, 2019.
4. Edwin, M., Saranya Nair, M., Joseph Sekhar, S., A comprehensive review for power production and economic feasibility on hybrid energy systems for remote communities. *International Journal of Ambient Energy*, DOI: 10.1080/01430750.2020.1712252, 2020.
5. Nehrir, MH., Wang, C., Strunz, K., Aki, H., Ramakumar, R., Bing, J., Miao, Z., Salameh, Z., A Review of Hybrid Renewable/Alternative Energy Systems for Electric Power Generation: Configurations, Control, and Applications. *IEEE Transactions of Sustainable Energy*, 2 (4), 392-403, 2011.
6. Iniyan, S., Sumathy, K., An optimal renewable energy model for various end-uses. *Energy*, 25, 563-575, 2000.
7. Andriy Redko, Oleksandr Redko and Ronald DiPippo., Hybrid systems with renewable energy sources. In: *Low-Temperature Energy Systems with Applications of Renewable Energy*, pp. 289-328, Academic Press, 2020. DOI: 10.1016/B978-0-12-816249-1.00008-X.
8. Singal, SK., Varun, Singh, RP., Rural electrification of a remote island by renewable energy sources. *Renewable Energy*, 32, 2491-2501, 2007.
9. Craig, G., Eilin, W., Kevin, PM., Philip, O., Regional integration of renewable energy systems in Ireland – The role of hybrid energy systems for small communities. *Electrical Power and Energy Systems*, 44, 713-720, 2013.
10. Alvares, SG., Trepp, TH., Simulation of a solar driven aqua ammonia absorption refrigeration system Part 1: mathematical description and system optimization. *International Journal of Refrigeration*, 10, 40-48, 1987.

11. Boonrit Prasartkaew, Kumar, S., Experimental study on the performance of a solar-biomass hybrid air-conditioning system. *Renewable Energy*, 57, 86-93, 2013.
12. Dhanuskodi, S., Sukumaran, R., Wilson, VH., Investigation of Solar Biomass Hybrid System for Drying Cashew. *International Journal of ChemTech Research*, 5 (2), 1076-1082, 2013.
13. Nixon, JD., Dey, PK., Davies, PA., The feasibility of hybrid solar-biomass power plants in India. *Energy*, 56, 541-554, 2012.
14. Perez-Navarro, A., Alfonso, D., Alvarez, F., Ibanez, C., Sanchez, I., Segura, C., Hybrid biomass-wind power plant for reliable energy generation. *Renewable Energy*, 35, 1436-1443, 2010.
15. Ajai Gupta, Saini, RP., Sharma, MP., Modelling of hybrid energy system- Part I: Problem formulation and model development. *Renewable Energy*, 36, 459-465, 2011.
16. Shezan, SKA., Al-Mamoon, A., Ping, HW., Performance Investigation of an Advanced Hybrid Renewable Energy System in Indonesia. *Environmental Progress & Sustainable Energy*, 37 (4), 1424-1432, 2018.
17. Kumaravel, S., Ashok, S., An Optimal Stand-Alone Biomass/Solar-PV/Pico-Hydel Hybrid Energy System for Remote Rural Area Electrification of Isolated Village in Western-Ghats Region of India. *International Journal of Green Energy*, 9, 398-408, 2012.
18. Belgin Emre Turkay, Ali Yasin Telli, Economic analysis of standalone and grid connected hybrid energy systems. *Renewable Energy*, 36, 1931-1943, 2011.
19. Elkadeem, MR., Shaorong Wang., Swellam W, Sharshir., Eman G, Atia., Feasibility analysis and techno-economic design of grid-isolated hybrid renewable energy system for electrification of agriculture and irrigation area: A case study in Dongola, Sudan. *Energy Conversion and Management*, 196, 1453-1478, 2019.
20. Sima Ashrafi Goudarzi, Farivar Fazelpour, Gevork B, Gharehpetian, Marc A. Rosen, Techno-Economic Assessment of Hybrid Renewable Resources for a Residential Building in Tehran. *Environmental Progress & Sustainable Energy*, 38 (5), 13209, 2019.
21. Murugaperumal, K., Ajay, P., Vimal Raj, D., Feasibility design and techno-economic analysis of hybrid renewable energy system for rural electrification. *Solar Energy*, 188, 1068-1083, 2019.
22. Edwin, M., Joseph Sekhar, S., Thermo-economic assessment of hybrid renewable energy based cooling system for food preservation in hilly terrain. *Renewable Energy*, 87, 493-500, 2016.
23. Edwin, M., Joseph Sekhar, S., Techno-economic studies on hybrid energy based cooling system for milk preservation in isolated regions. *Energy Conversion and Management*, 86, 1023-1030, 2014.
24. Shaopeng Guo, Qibin Liu, Jie Sun, Hongguang Jin, A review on the utilization of hybrid renewable energy. *Renewable and Sustainable Energy Reviews*, 91, 1121-1147, 2018.

25. Mehdi Bagheri, Navid Shirzadi, Elahe Bazdar, Christopher A, Kennedy, Optimal planning of hybrid renewable energy infrastructure for urban sustainability: Green Vancouver. *Renewable and Sustainable Energy Reviews*, 95, 254-264, 2018.
26. Vendoti Suresh, Muralidhar, M., Kiranmayi, R., Modelling and optimization of an off-grid hybrid renewable energy system for electrification in a rural areas. *Energy Reports*, 6, 594-604, 2020.
27. Abhishek Tiwary, Stanislava Spasova, Ian D. Williams, A community-scale hybrid energy system integrating biomass for localised solid waste and renewable energy solution: Evaluations in UK and Bulgaria. *Renewable Energy*, 139, 960-967, 2019.
28. Jijian Lian, Yusheng Zhang, Chao Ma, Yang Yang, Evance Chaima, A review on recent sizing methodologies of hybrid renewable energy systems. *Energy Conversion and Management*, 199, 112027, 2019.
29. Edwin, M., Joseph Sekhar, S., Techno-Economic evaluation of milk chilling unit retrofitted with hybrid renewable energy system in coastal province. *Energy*, 151, 66-78, 2018.
30. Edwin, M., Saranya Nair, M., Joseph Sekhar, S., Hybrid Energy-Based Chilling System for Food Preservation in Remote Areas, in: Innovation in *Energy Systems - New Technologies for Changing Paradigms*, Taha Selim Ustun (Ed.), IntechOpen, London, 2019.
31. Carrie, H., Samuel, G., Jonathan, W., Evaluation of energy efficiency and renewable energy generation opportunities for small scale dairy farms: A case study in Prince Edward Island, Canada. *Renewable Energy*, 67, 20-29, 2014.
32. Beno Wincy, W., Edwin, M., Techno-economic assessment on the implementation of biomass gasifier in conventional parboiling rice mills. *International Journal of Energy Research*, 44, 1709-1723, 2020.

2

Implementation of Hybrid Renewable Energy Projects in Rural India—A Case Study

Utpal Goswami[1]* and Arvind Kumar[2]

[1]West Bengal Renewable Energy Development Agency, West Bengal, India
[2]Consultant, The World Bank-MNRE-SBI Technical Assistance Programme, New Delhi, India

Abstract

The major objectives of this chapter are to establish a hybrid microgrid system to provide an uninterrupted power supply in rural areas. Such a hybrid system would include different sources of renewable energy like biogas, solar photovoltaic (PV), wind power, and a battery storage unit. Considering this, a case study has been performed by installing a hybrid microgrid project at Baliara, Mousuni Island, South 24 Parganas, Sundarban, West Bengal, India. The outcome of the study reveals the techno-economic variability the aforesaid pilot project.

Keywords: Hybrid, microgrid, solar, wind, biogas

2.1 Introduction

An increase in the demand for energy is an indicator of a country's growth and progress. In India, however, rural areas cover a major portion of the geography. These areas either lack access to electricity or have grid connection that is unstable or weak. The government of India had set a target for ensuring that every part of the country would be connected to the national grid by 2025. The government of India had also set a target date of March 2021 for ensuring that every domestic consumer has access to twenty-four hours of electric power [1]. However, the inadequacy of the distribution and transmission system grid infrastructure in rural areas was one of the

**Corresponding author*: utpal_cgr@rediffmail.com

Umakanta Sahoo (ed.) Hybrid Renewable Energy Systems, (37–62) © 2021 Scrivener Publishing LLC

major hindrances in achieving this goal. Distributed sources of renewable energy can be projected as a potential solution to help in matching the demand and supply of energy. Owing to the fact that the capacity utilization factor of solar PV or wind power is not high, neither of them can demonstrate dependable performance for meeting the demand of continuous load. To overcome this challenge, it is important to mix energy produced from bio-degradable waste and animal manure with these energy sources and a battery system for storing this energy (BESS). Thus the option that might prove to be effective in providing electrification in rural areas is a combination of BESS and renewable energy sources that is intelligent and targeted. For a particular load pattern, such a combination would have a distinct network and form a hybrid microgrid. An equivalent method of modeling that improves the use and reliability of renewable energy sources in the power system was proposed in their work by Caiet *et al.* [2]. According to the proposal, the model would ensure the microgrid's dynamic response and improve efficiency. Olivares *et al.* [3] discussed the limitation and challenges of renewable energy source microgrid reliability. A microgrid's different aspects of operation and strategies for control were also discussed by the authors. The future of distribution network was proposed by Nikmehr and Ravadanegh [4] through an explanation of the Multi-Microgrid System (MMGs). The economic analysis on minimum cost base was done among the main distribution and microgrid to achieve economic optimization. For minimization of the cost of power systems network in future, it was proposed that power should be shared between the main grid and microgrid. Rocabert *et al.* [5] discussed the management of microgrid for the scattered sources of energy having a connection to the main network. It was shown that through the implementation of this management strategy, there were improvements in terms of efficiency and stability of voltage. A test bed of microgrids was introduced by Hossain *et al.* [6]. The researchers also conducted a survey of different microgridsthe world over. Ma *et al.* [7] discussed the challenges that are faced while trying to integrate renewable energy sources within a microgrid. The researchers made use of the predictive control method model. Using this model simulation they were able to take a decision about the optimized way to share power. Operational constants and minimization of the cost of capital were considered in the simulation. By developing different microgrid simulation models using Particle Swarm Optimization (PSO) tool, the different models of microgrid such as Islanded mode and the grid-connected mode were discussed in detail by Chung *et al.* [8]. The challenges of controlling and managing power of different micro sources in a hybrid system (DC and AC) were discussed by Eghtedarpour and Farjah [9]. They proposed a decentralized method in which there would be no requirement for the communication between the different sources of energy. The PSCAD simulation was used for validating the performance of such a system. In order to ensure that the microgrid operates optimally, Shi *et al.* [10] discussed a system of energy management for the different energy sources like wind, solar, etc., along with a system for storing energy. The researchers proposed a method of sharing power among the different generators so that there is optimal scheduling between the central and local microgrid controllers. Carlos *et al.* [11] discussed the technique of minimizing fuel of different microgrid sources. They have also proposed a method to minimize cost. A sustainable microgrid design made of diesel generator (DG), PV, pumped hydro storage, solar PV, and battery storage was proposed by Kumar *et al.* [12]. The different sources and storage of renewable energy along with the growth of load on a daily, monthly and yearly basis was also presented by the researchers. For meeting the demand for load power with stability and reliability of DC bus voltage, a method of balance and control of power for a DC hybrid microgrid system was proposed by Changjie *et al.* [13]. González *et al.* [14] proposed a remote monitoring unit platform for the source of renewable energy with a smart microgrid that is based on hydrogen. Hossam *et al.* [15] described the use of static VAR compensator under islanded and grid-connected condition for the control of an AC-DC hybrid microgrid. In the literature [16–18] there have been proposals for control and power management of different hybrid microgrid. Environmental and economic scheduling of smart home applications using microgrid was discussed by Zhang *et al.* [19].

Through the above literature review, it can be concluded that optimization of capacity selection and distributed energy source integration along with proper BESS would ensure that there is an uninterrupted supply of power to the consumer even when there is either a weak or no presence of grid. However, hardly any of the researchers have pointed to the integration of biomass energy with wind, solar, and BESS for the operation and designing of a microgrid. Experimental study, along with simulation, has been used in this work for verifying the capability of a distributed renewable energy microgrid (inclusive of biomass) to share power so as to ensure that there is reliability of power supply. India has shown significant growth in implementing renewable energy sources based microgrid throughout the nation in the past several years through central and regional energy implementing agencies of the government of India and private sectors. In most of the cases, microgrids have been implemented with isolated distributed generators (DGs) of coordinated operation within it. A hybrid/tribrid system with multiple renewable energy sources like wind, solar, biogas and battery storage system for the purpose of 24hrs uninterrupted power supply for the no grid or weak grid regions of rural sectors of India.

2.2 Overview of Microgrid

A local grid that has the capability of accommodating multiple sources of renewable energy is referred to as a tribrid microgrid. Such a grid either operates in islanded or in grid-connected mode. The resilience of the grid is enhanced as a result. It can also aid in fast recovery of the grid in the event of a grid fault. Such microgrids are therefore of critical importance in remote places where the distribution grid is either absent or too weak to be able to handle critical load. Some of the highlighted features of such a hybrid microgrid are as follows [20]:

1. A multitude of different sources of renewable energy connected in a network of distribution along with controllable loads.
2. The mode of operation can be either off or on-grid.
3. For the management of source and demand side, there is a centralized controller that is intelligent.

2.3 Basic Structure of Hybrid System

Main bus bars along with the respective converters are used to connect the sources of renewable energy like biogas, wind, solar, etc. The point of common coupling acts as the connection between the micro and the main grid. Power flows to the connected loads from the generator. The basic model of the microgrid is illustrated in Figure 2.1.

There are mainly two modes in which a microgrid can operate:

(i) **Connected with the grid**: The microgrid is joined with the local grid in this mode. In this mode, the microgrid draws power from the low-voltage local grid in case of deficit and supplies power to the grid in event of excess power.

(ii) **Off-Grid:** This mode is also referred to as islanded mode. In this mode, the microgrid can provide power to the load continuously without interruption when there is no grid or the local grid has developed a fault.

For the smooth operation of the microgrid its technical and commercial viability needs to be ensured. For ensuring the commercial or economic viability of the project it is necessary to minimize O&M and Capital cost while maximizing the revenue. Minimization of losses while maximizing

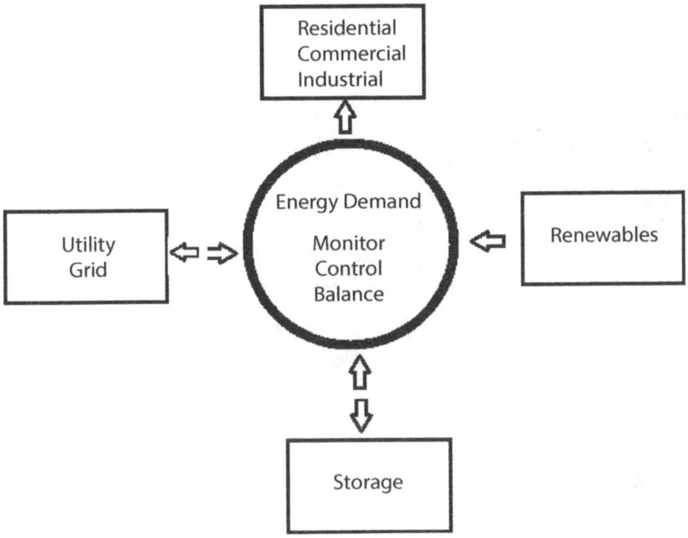

Figure 2.1 Basic Microgrid structure.

the efficiency is required to ensure technical viability. Some other technical aspects that need to be paid attention to are load balancing, fluctuation of voltage and regulating the frequency [21].

2.4 Hybrid Microgrid Control

Following are the components of a microgrid control [22]

i. Controller of micro source (MSC)
ii. Controller of load (LC)
iii. Controller of distribution system (DSC)
iv. Controller of Microgrid (MGC)

Every source of renewable energy having a connection in the microgrid is monitored and controlled by the MSC. Along with sending information from each of the sources of renewable energy to MGC, the MSC also ensures the management of battery. Along with controlling frequency and voltage the MGC also manages the curtailment and consumption of load. Additionally, the MGC ensures connection of the micro sources by acting as an interface. The LCs and MSCs are also synchronized by the MGC. The decision of whether the microgrid will be operated in islanded

Figure 2.2 Project location for hybrid plant.

or grid-connected mode would be taken by DSC. DSC also ensures that the entire system remains protected.

2.5 Project Location

The site is aerially about 2.5 km to the east of Dhablat village and 3.17 km to the southeast of Beguakhali village;it is located about 1.10 km from the coastline. The proposed location is represented by the 25 m high Ganga Sagar wind monitoring station. The wind Power Generating Station station was located close to Kapil Muni Ashram Mandir. Project location of hybrid power plant is marked in yellow in Figure 2.2.

2.6 Load Profile Study of Proposed Location

The microgrid's load is considered as 100 households. The average consumption of each house is around 3 kW. Average unit consumption in each household is around 9 units. In 2019, the consumption of power on a daily average and yearly average basis is given in the following Figure 2.3 and

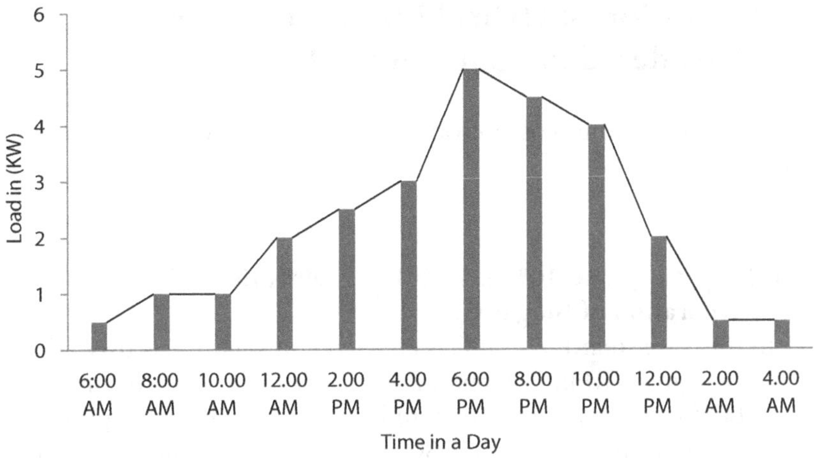

Figure 2.3 Average load pattern of household in April.

Figure 2.4. It was observed that the highest load requirement comes in April around 310 kW for 100 households. The highest load in April is around 5 kW, observed between 5PM and 8 PM. Our proposed hybrid microgrid system must be capable to supply the highest load demand and total load demand to ensure reliability.

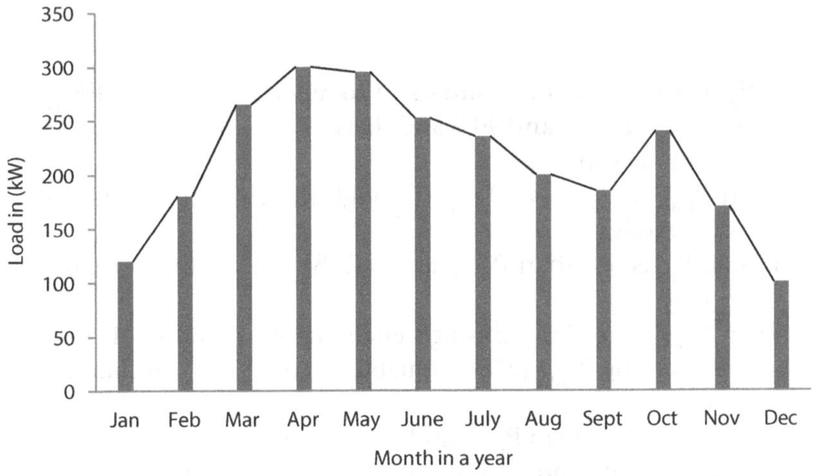

Figure 2.4 Average demand of power on a yearly basis.

2.7 Operation of Hybrid Microgrid System Considered for Current Study

For microgrid's optimum operation, the algorithm for energy management [23] is given below:

Case 1:

If **$P_{LOAD} < P_{PV}$ and 10% ≤ $VRFB_{SOC}$ ≥ 90%, the isolated mode generation of Biogas Gen.**

In this case solar PV's MSC is enabled and solar PV is used for charging VRFB. The charging of VRFB is done after using LC for satisfying load demand (P_{LOAD}) and the extra power is exported to grid. The DSC is used for delivering power to grid.

Else if, 90%≥ $VRFB_{SOC}$ and $P_{LOAD} < P_{PV}$ with grid condition that is stable,

Then $P_{PV} \longrightarrow P_{LOAD} + P_{GRID}$ (by enabling MSC VRFB is not in operation at this moment)

Else if, 90% ≤ $VRFB_{SOC}$ and $P_{LOAD} < P_{PV}$ with no grid or weak grid condition,

Then $P_{PV} \longrightarrow P_{LOAD}$ with grid that is in isolated or islanded mode and for dump loads like pump, excess power is used (by enabling DSC for microgrid's islanding from main grid and for control of load side and source MSC and LC is used).

Case 2:

If **P_{PV} is less than or not sufficient to meet P_{LOAD} and $VRFB_{SOC}$ is between 10% and 90% and there is sufficient P_{Biogas}** (>30% of its total capacity),

Then $P_{Biogas} + P_{PV} \longrightarrow P_{LOAD}$ (through the enabling of LC and MSC mode)

Else if, P_{PV} is less than P_{LOAD} and $VRFB_{SOC}$ is less than or equal to 10%,

Then $P_{Biogas} \longrightarrow P_{LOAD}$ through enabling of LC and P_{PV} is used for charging P_{VRFB} (MSC is enabled to put VRFB on charging mode)

Else if, P_{PV} is less than P_{LOAD} and P_{Biogas} is also less than 30% of its total capacity and $VRFB_{SOC}$ is between 10% and 90%;

Then the P_{PV} and P_{VRFB} is used for meeting the demand of power by load (P_{LOAD}) by making the LC enabled.

Case 3 (At night or cloudy weather):

If, P_{PV} is muchless than P_{LOAD} and P_{Biogas} is more than P_{load} while $VRFB_{SOC}$ is less than or equal to 10% and there is stable grid,

Then P_{Biogas} is used for meeting P_{LOAD}

If, P_{PV} is very less than P_{LOAD} and P_{Biogas} is more than P_{LOAD} whereas $VRFB_{SOC}$ is less than or equal to 10% and there is either no grid or weak grid,

Then P_{Biogas} is used for meeting P_{LOAD} (DSC is enabled for working in islanded mode from grid)

Else if, P_{PV} is muchless than P_{LOAD} and P_{Biogas} is also muchless thanP_{LOAD} with $VRFB_{SOC}$ also being less than 10% and grid is stable,

Then, P_{GRID} is used for meeting P_{LOAD} (by ensuring LC and DSC is enabled),

Else if P_{PV} is muchless than P_{LOAD} and P_{Biogas} is less than P_{LOAD} with $VRFB_{SOC}$ more than or equal to 50% (Using VRFB for meeting the demand for energy over a long period).

Case 4 (At night or cloudy weather) Sufficient Wind Speed available:

If, P_{PV} is muchless than P_{LOAD} and P_{Biogas} is more than P_{load} while $VRFB_{SOC}$ is less than or equal to 10% and there is stable grid, Sufficient Wind Speed is available to charge Battery

Then P_{Biogas} is used for meeting P_{LOAD}

If, P_{PV} is muchless than P_{LOAD} and P_{Biogas} is more than P_{LOAD} whereas $VRFB_{SOC}$ is less than or equal to 10% and there is either no grid or weak grid,

Then P_{Biogas} is used for meeting P_{LOAD} (DSC is enabled for working in islanded mode from grid)

Else if, P_{PV} is much less than P_{LOAD} and P_{Biogas} is also much less thanP_{LOAD} with $VRFB_{SOC}$ also being less than 10% and grid is stable,

$VRFB_{SOC}$ is charged through Wind turbine generator,

Then, P_{GRID} is used for meeting P_{LOAD} (by ensuring LC and DSC is enabled).

Else if P_{PV} is much less than P_{LOAD} and P_{Biogas} is less than P_{LOAD} with $VRFB_{SOC}$ more than or equal to 50% (Using VRFB for meeting the demand for energy over a long period),

Then P_{Biogas} together with P_{VRFB} is used for meeting P_{LOAD} (by ensuring LC and MSC is enabled).

2.8 Technical Specification of Hybrid System

This project would aim to design and install a tribrid/hybrid system called RE tribrid microgrid at Baliara, Mousuni Island, South 24 Parganas, Sundarbon West Bengal, of cumulative capacity 550 kW. Out of total capacity 100 kWp solar PV, wind generator of 250 kW and biogas digester of 350 m^3 along with a gas engine of 150kVA (120 kW). Besides these, there would also be associated with piping and instrumentation, 50kW/30kWh VRF battery. The capacity estimation of solar, wind, biogas has been carried out by HOMER software. The output of the RE tribrid microgrid is connected with local low-voltage grid and loads with a 415V low-voltage distribution line. The purpose of the RE tribid microgrid system is to supply 24hrs power to the loads associated with Microgrid Centre. The whole system is synchronized with the local grid through a synchronizing panel and the metering, monitoring and control is done through Lab VIEW software which is located within the main control panel. The salient feature of the RE tribrid microgrid system/project is shown in Table 2.1.

2.9 Modeling of Hybrid Microgrid System

a. Photovoltaic System Modeling:

The estimation of a PV generator's DC power is given by,

$$P_{pv\_dc} = I_{n\_eff} * \eta_g * A_g \tag{2.1}$$

Where,

I_{n_eff} = Effective incident irradiance

η_g = Efficiency of PV generator

A_g = Effective surface area of PV generator

The generated power by PV array at STC is illustrated as,

$$P_{pv\_peak} = I_n^* * \eta_g^* * A_g \tag{2.2}$$

Where,

P_{pv_peak} = PV array's peak DC power

$$I_n^* = 1000 \frac{W}{m^2}$$

Table 2.1 Salient features of the RE hybrid microgrid system project.

1	**Project site**	Microgrid Mousuni Island
2	**Location**	21.65°deg N & 88.21°deg E
3	**Irradiation and wind speed considered**	SRRA station, Mousuni Island
4	**Type of system**	RE tribrid Microgrid
5	**Type of technologies considered**	Solar PV, wind, biogas and VRF battery storage
6	**Proposed capacity**	555 kW Hybrid system
7	**Capacity of each system proposed**	
8	**Solar PV**	200kWp
9	**Wind**	250kW
10	**Biogas**	120 kW
11	**VRF Battery**	50kW/30 kWh (additional)
12	**Projected energy production per year**	0.65MU/Year
13	**Total project cost**	Rs. 2,50,00,000.00
14	**Project owner and funding authority**	Govt. of West Bengal
15	**Name of utility company**	WBSEDCL
16	**Approved tariff**	Net Metering
17	**Expected life of power plant**	25 years

From equation (2.2); If we substitute A_g 's value in equation (2.1), then:

$$P_{pv\_dc} = P_{pv\_peak} * \frac{I_{n\_eff}}{I_n^*} * \frac{\eta_g}{\eta_g^*} \tag{2.3}$$

The thermal loss of the module due to operational cell temperature is defined by the ratio $\frac{\eta_g}{\eta_g^*}$. Another way of expressing this is:

$$\frac{\eta_g}{\eta_g^*} = 1 - \beta(T_c - T_c^*) \tag{2.4}$$

Where,

$$T_c = T_a + C_T * I_{n\_eff} \tag{2.5}$$

$$C_T = \frac{NOCT - 20}{0.8 * I_n^*} \tag{2.6}$$

$$T_c^* = 25°C$$

The ambient temperature is denoted by T_a

$$\beta = \text{Coefficient of temperature loss}$$

$$NOCT = \text{The cell temperature for nominal operation}$$

The manufacturer of the module gives the value of both β and for simulating the 10kWp solar PV system.

b. Wind Generator Modeling:
A wind generator's output power is primarily dependent on its power output curve, hub of the mast's height and wind speed's distribution function at a particular location.

A wind turbine's power output can be illustrated as,

$$P_w(v) = P_R * \frac{v - v_C}{v_R - v_C}(v_C << v << v_R) \tag{2.7}$$

$$= P_R\,(v_R << v << v_F)$$

$$= 0\ (v << v_C \text{ and } v >> v_F)$$

Where,

P_R = electrical power's rated output
v_C = cut-in speed of wind
v_R = Speed of wind as per rating
v_F = cut-off speed of wind

The speed of the wind and the power generated by wind turbine is significantly affected by the variation in the height of turbine hub. For estimation of vertical wind profile over horizontal contour the following mathematical model is used

$$\frac{V_n}{V_o} = \left(\frac{H_n}{H_o}\right)^{\propto} \tag{2.8}$$

Where,

V_n = Speed of wind at new elevation point
V_o = Speed of wind at original elevation point
H_n = New elevation
H_o = Original elevation
$\propto$ = Exponent of power law

A number of parameters like, time of day, elevation, season, temperature etc. have an effect on the value of α [24].

c. Biogas plant Modeling:

The equation of power generation from the biogas plant [136] is

$$P_{bg} = \eta_{gen} * D_w * Y_{bg} \tag{2.9}$$

Where,

η_{gen} = denotes the efficiency of the generator
D_w = per day collection of bio-degradable table and kitchen waste (kg)
Y_{bg} = The biogas generated from 1 kg of bio-degradable waste

The equation for calculating the digester's volume is:

$$V_{dg} = (V_{slr} * T_{rtn})/0.5 \tag{2.10}$$

Where,

V_{slr} = The slurry volume generated in a day. It can also be written as

$$V_{slr} = 2 * D_w \tag{2.11}$$

For producing the slurry, water is mixed in the ratio of 1:1

$$T_{rtn} = \text{The retention time of the slurry in digester}$$

Of the total volume of the digester, 50% is occupied by slurry and the rest is occupied by generated gas. There are two main parts of the gas generator – the synchronous generator and the internal combustion engine. The IC engine also has attached with it a governor for adjusting the supply of the fuel and thus controlling the speed of the shaft. The PID controller controls the governor. The synchronous generator has attached with it a voltage regulator and an exciter. The generator's terminal voltage is controlled by the voltage regulator. For maintaining the reliability of the load and maximum fuel efficiency, it is required that the gas engine is run at the highest level of load. The gas engine's rated power varies between 0 and max load demand. The following equation expresses this condition

$$0 \leq P_{Rated} \leq P_{Load\_max} \tag{2.12}$$

Where,

P_{Rated} = Gas engine's rated power
P_{Load_max} = At load peak power

The quantity of fuel that a gas generator takes for driving a load is

$$F_{gas}(t) = (m_0 * P_{gas}(t) + n_0 * P_{Rated}) \tag{2.13}$$

Where,

$F_{gas}(t)$ = Consumption of fuel in m^3/hour
$P_{gas}(t)$ = Power of the generator at time't'
$m_0 \& n_0$ = Fuel's calibration coefficient

The consumption of fuel by a gas generator depends on rated power as well as power generated and is described by the above equation (equation 2.13). The consumption of fuel by a gas generator is thus limited to,

$$P_{gas\_min}(t) \leq P_{gas}(t) \leq P_{gas\_max}(t) \tag{2.14}$$

Where,

$P_{gas\_min}(t)$ = Operating power's minimum value
$P_{gas\_max}(t)$ = Operating power's maximum value

d. VRFB Storage's Modeling:

Some of the advantages of VRFB over other forms of conventional batteries are: highest cycle time (≈13000), capacity to store energy and scalability of power. Thus for ensuring reliability and security of energy for renewable energy applications that are stationary, VRFB is the most potential storage. The primary parameters to be considered for ensuring proper size of the battery bank are operating temperature, wind and solar power variations and characteristics for charge/discharge. In SOC terms the expression of voltage is.

$$V_{bat} = V_{stack}(SOC, Q, n_{cell}) \pm I_{stack} * R_{specific}(I_{stack}, Q, T) \quad (2.15)$$

Where,

V_{stack} = open circuit Stack voltage (V)
I_{stack} = Stack current (A)
$R_{specific}$ = Specific stack resistance (mΩ)
Q = Rate of flow (ml/sec.)
T = Temperature of operations

The formula for state of charge is

$$SOC = \frac{C(t)}{C_{Ref}} \quad (2.16)$$

Where,

$C(t)$ = The capacity of battery at time instant't'
C_{Ref} = Reference capacity

Now, the formula for $C(t)$ is,

$$C(t) = q(t_0) + q_c(t) + q_d(t) \quad (2.17)$$

Where,

$q(t_0)$ = Quantity of charge at beginning
$q_c(t)$ = The amount of charge getting in during the process of charging
$q_d(t)$ = The amount of charge getting out during the process of discharging

The battery's state of health (SOH) which is a very important parameter is defined as,

$$SOH = \frac{C_{Ref}(t)}{C_{Ref\_nom}} \tag{2.18}$$

Where,

C_{Ref_nom} = Nominal capacity of reference available

The estimation of new capacity reference is,

$$C_{Ref}(t) = C_{Ref}(t - \Delta t) - \Delta C_{Ref}(t) \tag{2.19}$$

The formula for capacity loss is,

$$-\Delta C_{Ref}(t) = C_{Ref\_nom} * Z * (SOC(t - \Delta t) - SOC(t)) \tag{2.20}$$

The new equation for SOH is,

$$SOH(t) = \frac{C_{Ref}(t - \Delta t)}{C_{Ref\_nom}} - Z * (SOC(t - \Delta t) - SOC(t)) \tag{2.21}$$

Where,

Z = Coefficient of linear aging.

For the complete system optimization, properly sizing the components of the subsystem and modeling of individual components is important. The model's simulation is illustrated in the following equation:

$$P_{mg} = P_{bg}(t, Q) + P_{pv}(t, I, T_a) + P_w(t, v) + P_b(t, SOC) \tag{2.22}$$

Where,

P_{mg} = Microgrid's total power (W)
P_{bg} = Amount of power that the biogas plant generates (W)
P_{pv} = Amount of power that solar plant generates (W)
P_w = Amount of power that wind turbine generates (W)
P_b = Amount of power that battery bank generates (W)
v = Velocity of Wind (m/sec.)

2.10 Last One Year Output of Hybrid Microgrid Plant

The load demand and the combined power generation (VRFB, Biogas, wind and Solar PV) is illustrated in the form of a clustered stacked bar chart as shown in Figure 2.5. As can be seen from the figure for every hour of the day the load demand (illustrated by the off-blue bar) is met by the combined power generated. Thus the microgrid is able to ensure that the load has 24X 7 accesses to electrical power.

The amount of energy produced by different sources of renewable energy and the percentage of power shared by each source for serving the load is given in the following Table 2.2. As can be seen from Table 2.2, the amount of power generated from wind is negligible as compared to other sources.

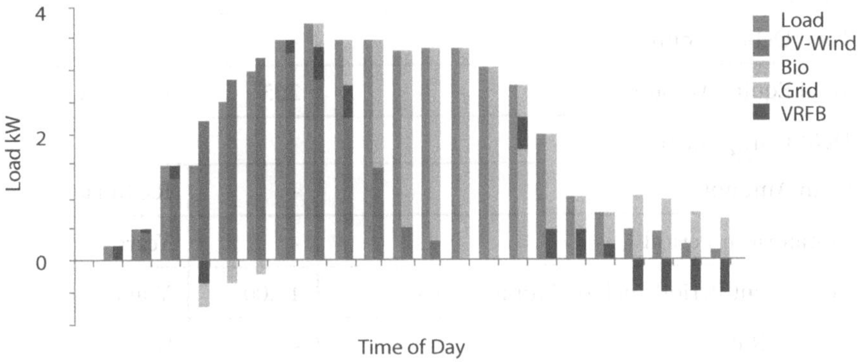

Figure 2.5 Peak load that the hybrid micro-grid caters to over a day.

Table 2.2 Percentage energy share by different hybrid microgrid sources.

Hybrid microgrid sources	% Sharing
Solar	46
Wind	20
Bio Gas	25
Battery	9

Table 2.3 Basic assumptions for cost analysis.

Project particulars	**Data**	**Unit**
Installed Power Generation Capacity	550	kW
Auxiliary Power Consumption	0	%
Capacity Utilization Factor	17.6	%
Useful Life	25	Years
Tariff Period	25	Years
Power Plant Cost	265	Rs. In lac
Debt: Equity		
Debt	-	%
Equity	100.00	%
Total Debt Amount	-	Rs. In lac
Total Equity Amount	265	Rs. In lac
Debt Component		
Loan Amount	-	Rs. In lac
Moratorium Period	-	Years
Repayment Period(include Moratorium)	12.00	Years
Interest Rate	-	%
Equity Component		
Equity amount	265	Rs. In lac
Return on Equity for first 10 years	-	% p.a
Return on Equity 11th year onwards	-	% p.a
Weighted average of ROE	0	% p.a
Discounting Factor	10.64	%
Income Tax	33.99	%
MAT Rate (for first 10 years)	14	%
Depreciation		

(Continued)

Table 2.3 Basic assumptions for cost analysis. (*Continued*)

Depreciation Rate for first 12 years	5.83	%
Depreciation Rate 13th year onwards	1.54	%
Working Capital		
O&M Charges	1	Months
Maintenance Spares	15.00%	% of O&M expenses
Receivables for Debtors (actual Fixed cost)	2	Months
Interest of Working Capital	13.26	%
O & M Expenses/Year	7.95	Rs. In lac
O&M Escalation	5.72	%

This is because as compared to other regions the profile of wind power is weakest in this region while the power generated from solar energy is highest among all the sources. Thus, the baseload of this hybrid system is solar PV.

2.11 Financial Analysis

a. Major Assumption

The estimated project cost has been worked out on the following basis and assumptions: major assumptions for the preparation of the cost estimate and financial analysis for the Wind Solar Hybrid Grid-Connected Solar PV power Plant are summarized below in Table 2.3.

b. Capital Cost Calculation

The capital cost of the project including Solar, Wind, Biogas and battery storage are depicted in Table 2.4.

2.12 Tariff Calculation

Details of the financial analysis and Levelized Tariff calculation is shown in the Annexure-I. Summary cost is depicted in Table 2.5.

Annexure-I

LEVELISED TARIFF CALCULATION

Year	---→	1	2	3	4	5	6	7	8	9	10	11	12	13	14	15	16	17	18	19	20	21	22	23	24	25
Gross Generation (Solar+Wind)	MU	1.70	1.69	1.67	1.66	1.64	1.63	1.61	1.60	1.58	1.57	1.55	1.54	1.53	1.51	1.50	1.49	1.47	1.46	1.45	1.43	1.42	1.41	1.40	1.38	1.37

Fixed Cost																										
O & M Cost	Rs. in Lac	14.00	14.80	15.65	16.54	17.49	18.49	19.55	20.66	21.85	23.10	24.42	25.81	27.29	28.85	30.50	32.25	34.09	36.04	38.10	40.28	42.59	45.02	47.60	50.32	53.20
Depreciation Rate	%	5.83	5.83	5.83	5.83	5.83	5.83	5.83	5.83	5.83	5.83	5.83	5.83	1.54	1.54	1.54	1.54	1.54	1.54	1.54	1.54	1.54	1.54	1.54	1.54	1.54
Depreciation	Rs. in Lac	36.07	36.07	36.07	36.07	36.07	36.07	36.07	36.07	36.07	36.07	36.07	36.07	9.53	9.53	9.53	9.53	9.53	9.53	9.53	9.53	9.53	9.53	9.53	9.53	9.53

Working Capital																										
Maintenance	Rs. in Lac	1.17	1.23	1.30	1.38	1.46	1.54	1.63	1.72	1.82	1.92	2.03	2.15	2.27	2.40	2.54	2.69	2.84	3.00	3.18	3.36	3.55	3.75	3.97	4.19	4.43
Spares	Rs. in Lac	2.10	2.22	2.35	2.48	2.62	2.77	2.93	3.10	3.28	3.46	3.66	3.87	4.09	4.33	4.58	4.84	5.11	5.41	5.72	6.04	6.39	6.75	7.14	7.55	7.98
Receivables	Rs. in Lac	2.22	2.35	2.48	2.62	2.77	2.92	3.09	3.26	3.44	3.64	3.84	4.06	4.28	4.52	4.78	5.04	5.33	5.62	5.94	6.27	6.62	6.99	7.39	7.80	8.24
Interest on Working Capital	Rs. in Lac	0.73	0.77	0.81	0.86	0.91	0.96	1.01	1.07	1.13	1.20	1.26	1.34	1.41	1.49	1.58	1.67	1.76	1.86	1.97	2.08	2.20	2.32	2.45	2.59	2.74
Total Working Capital	Rs. in Lac	5.49	5.80	6.13	6.48	6.85	7.24	7.65	8.08	8.54	9.02	9.54	10.08	10.65	11.25	11.89	12.57	13.28	14.03	14.83	15.67	16.56	17.50	18.49	19.54	20.65
Total Fixed Cost	Rs. in Lac	14.73	15.57	16.46	17.40	18.40	19.45	20.56	21.74	22.98	24.29	25.68	27.15	28.70	30.34	32.08	33.91	35.85	37.90	40.07	42.36	44.78	47.34	50.05	52.91	55.94

O & M Cost	Rs. per kWh	0.82	0.88	0.94	1.00	1.07	1.14	1.21	1.29	1.38	1.47	1.57	1.68	1.79	1.91	2.03	2.17	2.31	2.47	2.63	2.81	3.00	3.20	3.41	3.64	3.88
Depreciation	Rs. per kWh	2.12	2.14	2.16	2.18	2.20	2.22	2.24	2.26	2.28	2.30	2.32	2.34	0.62	0.63	0.64	0.64	0.65	0.65	0.66	0.66	0.67	0.68	0.68	0.69	0.69
Interest on Working Capital	Rs. per kWh	0.04	0.05	0.05	0.05	0.06	0.06	0.06	0.07	0.07	0.08	0.08	0.09	0.09	0.10	0.11	0.11	0.12	0.13	0.14	0.14	0.15	0.16	0.18	0.19	0.20

Total Cost of Generation	Rs. per kWh	0.87	0.92	0.99	1.05	1.12	1.20	1.28	1.36	1.45	1.55	1.65	1.76	1.88	2.01	2.14	2.28	2.43	2.60	2.77	2.95	3.15	3.36	3.59	3.82	4.08
Discounting Factor (10.64%)		1.00	0.90	0.82	0.74	0.67	0.60	0.54	0.49	0.44	0.40	0.36	0.33	0.30	0.27	0.24	0.22	0.20	0.18	0.16	0.15	0.13	0.12	0.11	0.10	0.09
Discounted Cost of Generation	Rs. per kWh	0.87	0.83	0.80	0.78	0.75	0.72	0.69	0.67	0.65	0.62	0.60	0.58	0.56	0.54	0.52	0.50	0.48	0.46	0.45	0.43	0.41	0.40	0.39	0.37	0.36
Levelized Tariff	Rs. per kWh	1.51																								

Table 2.4 Total capital cost of hybrid microgrid projects.

Sl. No.	Items	Price (in INR)
1	**PV Power Plant**	10040000.00
2	**Wind Power Plant**	7530000.00
3	**Biogas Power Plant**	6275000.00
4	**Battery Storage Plant**	1255000.00
5	**Power Evacuation up to Interconnection Point**	
5.2	Equipment & associate items for 415 V indoor switchyard LCR & MCR and indoor C&R Panel	8,00,000.00
5.3	DC Battery, Battery Charging Equipment and DCDB (for Power Distribution/accumulation AC side required	50,000.00
5.4	Power Evacuation from Solar Substation to existing 11/.415 kV Pole Mounted Substation with necessary protection system at termination	6,00,000.00
6	**Miscellaneous Items**	
6.1	Fire Protection System	50,000.00
6.2	Illumination system	30,000.00
6.3	Signage	5,000.00
6.4	Ventilation System	30,000.00
6.5	Air Conditioning System	50,000.00
6.6	Tool Kit & Measuring Instruments	10,000.00
6.7	Spares	29,157.50
6.8	CC TV and Surveillance System	30,000.00
7	**Hard Cost Excluding O&M (A)**	**2,14,52,381.00**
8	**Administrative Cost (5% of A) (B)**	**10,72,619.00**
9	**O&M Cost for 5 Years (C)**	**39,75,000.00**
10	**Total Project Cost (A+B+C)**	**2,65,00,000.00**

Table 2.5 Summary cost.

S. No. Sl. No.	Item	Amount
1.	**Hybrid Plant Capacity**	525 kWp Nominal
2.	**Project Cost**	
a)	Solar PV Power Plant	2.15 Cr.
c)	O&M Cost for five years	0.40Cr.
d)	Administrative Cost @5% of (a+b)	0.10 Cr.
3.	**Total Project Cost**	**2.65Cr.**
4.	**Govt. Fund**	**2.65 Cr. (100% of Project Cost)**
5.	**Levelized Tariff (for 25 Years)**	**Rs.1.50/kWh**

2.13 Conclusion

The real-time simulation has been performed to meet the 24/7 hour supply for local demand with and without grid condition. It has been found that 200 kwp, 250 kWp wind, and 120 kWp biogas generation is best suitable for this project. A smart hybrid microgrid comprising with storage, local-ensure 24/7 power supply in 100 households at Mousuni Island, South 24 Parganas, Sundarbon West Bengal. Owing to the renewable energy source's intermittent nature; there has been the implementation of an algorithm of energy management. The performance of the pilot projects reveals that hybrid microgrid systems successfully manage peak demand and continuous load demand successfully. Battery storage support is very rarely required under extreme contingency. The Levelized tariff for 25 years comes to around Rs.1.50 /kWh based on the consideration of initial investment amount and the revenue generated by the project seems to be viable from technical and financial perspective. The generalized solution proposed in this work is very effective in rural areas where there is either no grid or very weak presence of the grid. This would ensure zero loss of the probability of power supply at the consumer side.

References

1. Gandhi V, Bhasin P, Khanna D, Dhawan K, Seddon J, Beyond Off-Grid: Integration Mini- Grids With India's Evolving Electricity System. 2016. pp. 2-5.
2. Cai C, Liu H, Dai W, Deng Z, Zhang J, Deng L. Dynamic equivalent modeling of a grid-tied microgrid based on characteristic model and measurement data. Energies 2017;10:1–16.
3. Olivares DE, Mehrizi-Sani A, Etemadi AH, Cañizares CA, Iravani R, Kazerani M, *et al.* Trends in microgrid control. IEEE Trans Smart Grid 2014;5:1905–19.
4. Nikmehr Nima. Sajad najafi ravadanegh, optimal power dispatch of multi-micro-grids at future smart distribution grids. IEEE Trans Smart Grid 2015;6:1648–57.
5. Rocabert J, Azevedo G, Candela I, Teoderescu R, Rodriguez P, Etxebarria-Otadui I. Microgrid connection management based on an intelligent connection agent. IECON 2010;36:3028–33.
6. Hossain E, Kabalci E, Bayindir R, Perez R. Microgrid test beds around the world: state of art. Energy Convers Manage 2014;86:132–53.
7. Ma J, Yang F, Li Z, Qin SJ. A renewable energy integration application in a microgrid based on model predictive control. IEEE Power Energy Soc 2012:1–6.
8. Chung Y, Liu W, Cartes DA, Schoder K. Control parameter optimization for a microgrid system using particle swarm optimization. ICSET 2008.
9. Eghtedarpour N, Farjah E. Power control and management in a hybrid AC/DC microgrid. IEEE Trans Smart Grid 2014;5:1494–505.
10. Shi W, Xie X, Chu C, Gadh R. Distributed optimal energy management in micro-grids. IEEE Trans Smart Grid 2015;6:1137–46.
11. Carlos A, Aramburo H, Green TC, Mugniot N. Fuel consumption minimization of a microgrid. IEEE Trans Industry Appl 2005;41:673–81.
12. Kumar A, Singh AR, Den Y, He X, Kumar P, Bansal RC. Integrated assessment of a sustainable microgrid for a remote village in hilly region. Energy Convers Manage 2019;180:442–72.
13. Changjie Y, Hongwei W. Fabrice locment, manuela sechilariu, energy management of DC microgrid based on photovoltaic combined with diesel generator and supercapacitor. Energy Convers Manage 2017;132:14–27.
14. González I, Calderón AJ, Andújar JM. Novel remote monitoring platform for RES-hydrogen based smart microgrid. Energy Convers Manage 2017;148:489–505.
15. Gabbar Hossam A, Abdelsalam Abdelazeem A. Microgrid energy management in grid-connected and islanding modes based on SVC. Energy Convers Manage 2014;86:964–72.

16. Mehrasa M, Pouresmaeil E, Mehrjerdi H, Jørgensen NB, Catalão JPS. Control technique for enhancing the stable operation of distributed generation units within a microgrid. Energy Convers Manage 2015;97:362–73.
17. Reyasudin M, Khan B, Jidin R, Pasupuleti J. Multi-agent based distributed control architecture for microgrid energy management and optimization. Energy Convers Manage 2016;112:288–307.
18. Choudar A, Boukhetala D, Barkat S, Brucker J-M. A local energy management of a hybrid PV-storage based distributed generation for microgrids. Energy Convers Manage 2015;90:21–33.
19. Zhang D, Evangelisti S, Lettieri P, Papageorgiou LG. Economic and environmental scheduling of smart homes with microgrid: DER operation and electrical tasks. Energy Convers Manage 2016;110:113–24.
20. Rahmani-Andebili Mehdi. Dynamic and adaptive reconfiguration of electrical distribution system including renewables applying stochastic model predictive control. IET Gener Transm Distrib 2017;11(16):3912–21.
21. Rahmani-Andebili Mehdi. Adaptive, stochastic and dynamic control of energy storage systems integrated with renewable energy sources for power loss minimization. Renew Energy 2017. https://doi.org/10.1016/j.renene.2017.07.005.
22. Rahmani-Andebili Mehdi. Cooperative distributed energy scheduling in microgrids. Springer Nature Singapore Pte Ltd; 2018. p. 235–54.
23. Yang H, Lu L, Zhou W. A novel optimization sizing model for hybrid solar-wind power generation system. Sol Energy 2007;81:76–84.
24. Bhattacharjee A, Samanta H, Banerjee N, Saha H. Development and validation of a real time flow control integrated MPPT charger for solar PV applications of vanadium redox flow battery. Energy Convers Manage 2018;171:1449–62.

3

Techno-Economic Analysis of Hybrid Renewable Energy System with Energy Storage for Rural Electrification

Pradeep Kumar Sahu[1]*, Satyaranjan Jena[1] and Umakanta Sahoo[2]

[1]School of Electrical Engineering, KIIT, Deemed to be University, Bhubaneswar, India
[2]National Institute of Solar Energy, GOI, Haryana, India

Abstract
Hybrid renewable energy systems with energy storage are a suitable model for rural electrification. The hybrid systems' main sources are namely photovoltaic and wind, as these renewable resources are having very good potential in various regions of India. The energy storage, i.e., battery, has been used for uninterrupted power supply under the power outage due to seasonal variation or short transient of solar radiation. This study assesses some major issues and challenges in the various design of the hybrid systems. Moreover, the techno-economic feasibility analysis of various hybrid renewable energy systems for rural electrification has been investigated using HOMER software. A Comprehensive economic comparison among the various optimal configurations of hybrid renewable energy systems; such as wind/battery, PV/battery and wind/PV/battery is made on the basis of net present cost (NPC) and cost of energy (COE), and a feasible solution of a suitable configuration for rural electrification is recommended. It is also observed that the wind/PV/battery hybrid system has COE of 0.278 ($/kWh) and NPC of $454,242, which confirms that such combination is more economical for the remote location.

Keywords: Renewable power generation, hybrid energy system, energy management system, energy storage, techno-economic analysis

**Corresponding author*: pksahu.nitrkl@gmail.com

Umakanta Sahoo (ed.) Hybrid Renewable Energy Systems, (63–96) © 2021 Scrivener Publishing LLC

3.1 Introduction

In the present scenario, with increasing concern about fossil fuel exhaustion, the environmental aspects and increase in global energy demand, the development of renewable energy (RE) sources like Solar photovoltaic (PV), wind, micro-hydro, biomass, geothermal, ocean wave and tides have motivated many nations to provide innovative energy strategies. Among these sources, photovoltaic (PV) and wind turbines (WT) have gained worldwide attention due to their complementary nature, low operating cost, and modular nature [1, 2]. The hybrid renewable energy source (HES) system comprising multiple sources is commonly used for bulk power production. The RE source-based hybrid energy system is the most suitable solution for providing power to a rural location where the utility grid extension is not feasible. However, the main drawback of the RE source is that power generation is intermittent in nature. The active and reactive power could not be regulated easily as is done in conventional generation due to the variation of load. Therefore, power management between supply and demand may not be possible. The storage systems are integrated with the HES to ensure the reliable power supply and to look after the fluctuating nature of renewable sources. The power produced by the RE sources have different operating natures; it is, therefore, necessary to have a clearly defined and normalized framework for connecting these sources to frame an HES. Here, a local cluster of RE sources, energy storage, and loads are integrated together and capable of operating autonomously.

Vast research has been carried out in the field of economic viability assessment of numerous hybrid configurations. An extensive literature review [3] has been reported on the technical analysis and economic potential of various configurations of HES for telecommunication applications for Punjab India. The performance evaluation and techno-economic analysis of an isolated wind-PV hybrid energy system are addressed by Bhakta *et al.* [4] on an island of India. It is also observed that on the basis of net profit cost (NPC) and cost of energy (COE), the hybrid wind-PV system is more economical and reliable as compared to the sole PV system or sole wind energy system. A. Al-Sharaf *et al.* [5] have conducted a similar study for various sites in Saudi Arabia, taking into account five combinations of HES comprising wind energy system, PV system, battery storage system, hydrogen storage system and fuel cells. It is found that the PV/wind/batteries hybrid system is the most cost-effective one in comparison to other combinations.

3.2 HES Components

RE resources are available in abundance all across the world to fulfill our energy demand. However, the energy produced by most of these sources is unpredictable and fluctuating in nature. A single RE source also produces a relatively small amount of power as compared to conventional power plants. Hence, a hybrid system comprising multiple sources is commonly used in a remote location. These hybrid systems ensure more dependable and sustainable energy [6]. It also provides more economical operation than a single RE source [7]. The hybrid system shown in Figure 3.1 includes power production from RE sources, like Photovoltaic (PV), wind, micro-hydro, biogas, and geothermal, which have high potential in remote areas.

The HES is suitable for rural electrification where grid expansion is not feasible. The storage system is integrated with the HES to ensure a reliable power supply and to look after the fluctuating nature of renewable sources [8]. Research has been carried out for the last 50 years in renewable energy to facilitate clean and green energy [9]. Out of available energy sources, hydro and biogas are considered first-generation resources that have been commonly used for the past 100 years. Solar and wind, which are treated as second-generation, became popular because of developments in power converter and RE technologies [10]. Moreover, third-generation resources comprise geothermal, biomass, ocean, and

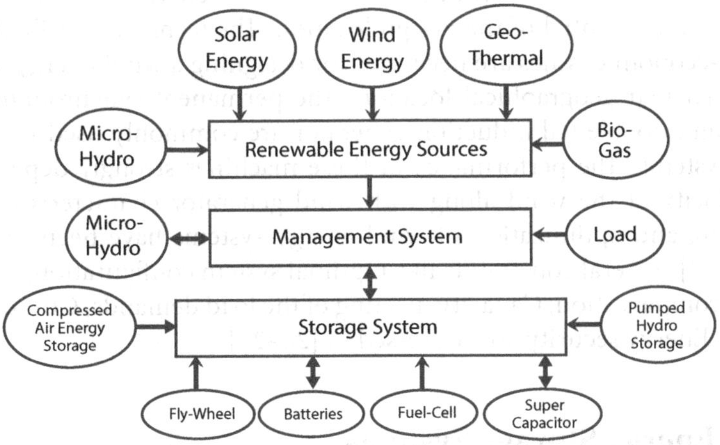

Figure 3.1 Hybrid renewable energy system.

tidal energy [11]. The evaluation of nanotechnology helps progress in research on these sources [12, 13]. Among the above-mentioned sources, solar photovoltaic and wind energy are the better options to produce energy on a large scale [14]. The hybrid system mainly consists of PV and wind with an energy storage system that is first reported in [9, 15] and further discussed in [16, 17].

A solar photovoltaic (PV) system produces electrical energy from solar energy. A solar cell is the smallest member of this system. Three classes of PV cells are available such as mono-crystalline, poly-crystalline, and thin-film, each possessing their own advantages and drawbacks [18]. Titanium oxide-coated PV cells are also developed nowadays due to the advancement in research in nanomaterials. Series and parallel combinations of solar cells are connected to form a module and then an array. The power produced by the PV system depends on external influences, like the surrounding temperature and solar irradiance, elevation angle of solar panels. The maximum power from the solar panel can be extracted by applying the maximum power point tracking (MPPT) algorithms. Various MPPT algorithms are proposed in the literature [19, 20]. Presently, Solar PV systems have the highest contribution in the power sector among all renewable energy sources. The leading countries like China (44 GW), Germany (38 GW), and Japan (35 GW) have the largest solar power installed capacities. The wind energy sources are the fastest developed renewable energy. The world's largest wind energy-producing countries are China with a capacity of 188 GW followed by USA (80 GW) and Germany (48 GW) [14]. The output power production of the wind energy system depends on the wind potential of the design location. The technical feasibility and techno-economic issues are necessary for designing a wind energy system in a particular geographical location. The permanent synchronous generator and double fed induction generator are commonly used to design these systems. The performance of these machines strongly depends on the velocity of the wind along with wind generator characteristics. The modeling and optimization of wind energy systems have been reported in [21–23]. Several constraints like Optimal system configuration, reactive power compensation, Capacity meeting of the load demands, Cost of wind energy, Energy security are addressed in [24–27].

3.3 Energy Storage Systems

The main drawback of the renewable sources is that power generation is intermittent in nature. The active and reactive power cannot be controlled easily as is done in the conventional generation due to the variation of load [28]. Therefore, due to the supply and demand gap, there may be a mismatch in power. Hence, to maintain the reliability of the HES system storage systems play a very important role especially in case of a stand-alone HES system. Due to the intermittent nature of RES, various power quality issues arise in the power system. In order to improve the power system quality and stability, storage is very much essential [29, 30].

The interim power to the load is supplied by the storage system when the RES is not able to match the load demand. Another advantage of the storage system is that the reactive power demand of the load is fulfilled at the time of islanding [31–33]. As per the literature, the various applications of the storage system are reported in [34]:

I. PQI (Power quality Improvement).
II. APS (Ancillary power supply).
III. STPS (Short-term power supply).
IV. PTLS (Peak time load sharing) during grid-connected system.

The various storage technologies available are discussed below. At present, the classification of power energy storage varies with different standards. According to the form of energy conversion, it can be divided into three categories, as given in the chart below.

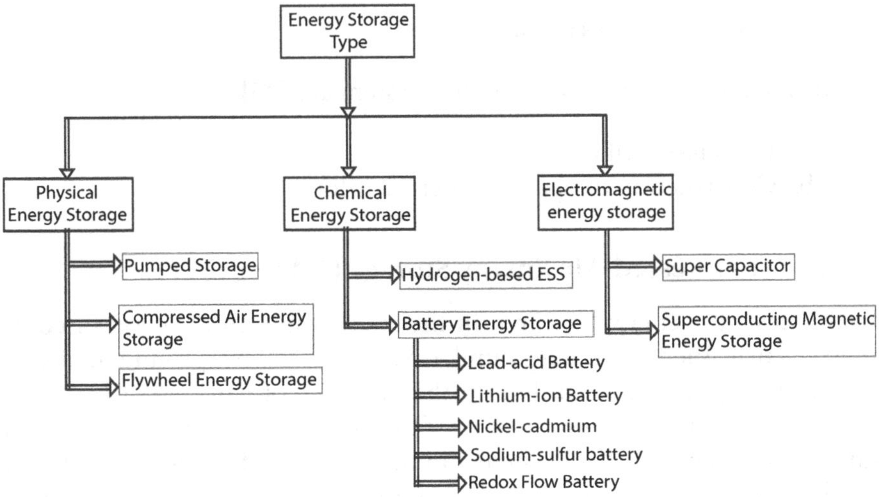

Figure 3.2 Energy storage type.

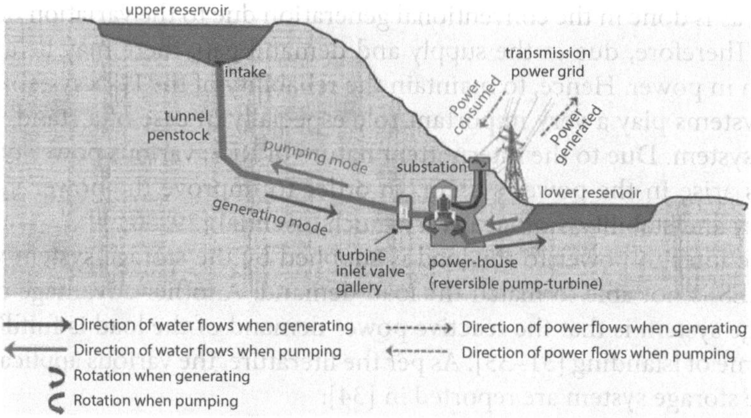

Figure 3.3 Pumped hydro storage [77].

3.3.1 Pumped Hydro Storage (PHS)

PHS generally is used as an energy storage system for power/grid applications as shown in Figure 3.3. This is widely implemented because of the simplest technology. In the PHS system, the water is pumped to a higher altitude during the availability of surplus energy and the same is utilized to generate power during the scarcity period. The advantages of this system are

a. Largest rating
b. Longer life period
c. Higher efficiency
d. Discharge losses are small.

The main disadvantages of the PHS system are [35]

a. It is site-specific
b. Construction to operation period is long

3.3.2 Compressed Air Energy Storage (CAES)

In this case, the energy is stored in the air chamber in the form of compressed air either in underground caverns or in overground tanks as shown in Figure 3.4. According to the principle of working, it is of two types. During the compression of air, if the heat released is wasted then it is known as a diabetic CAES system. On the other hand, if the released heat can be reused for decompression it is known as a adiabatic CAES system [36]. The merits of the CAES system are

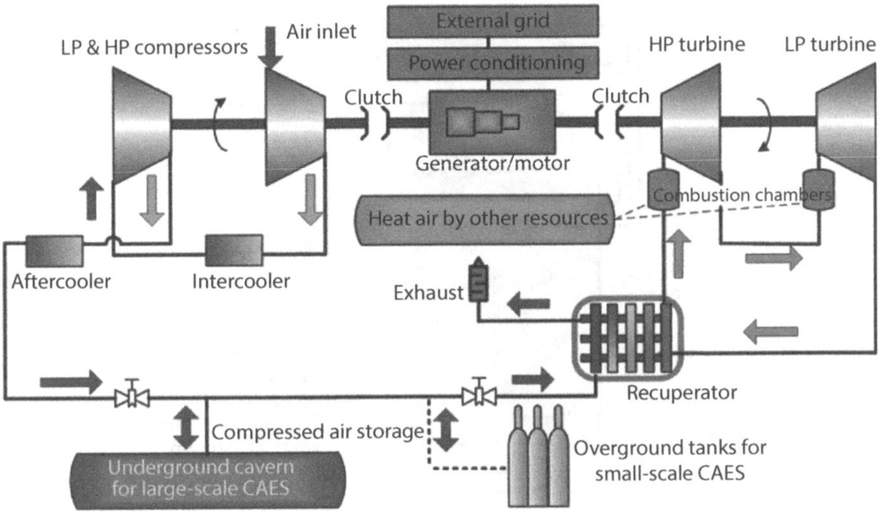

Figure 3.4 Compressed air energy storage [78].

a. High power rating
b. Longer life period
c. Discharge losses are small.

The main disadvantages of the CAES system is the same as the PHS system [30].

a. It is site-specific
b. Higher investment
c. Suitable for grid application

3.3.3 Flywheel Energy Storage (FES)

In flywheel, the energy is stored in the form of kinetic energy (Figure 3.5). An accelerated rotor that can rotate from 20,000 to 50,000 rpm is used to store the rotational energy. The rotor is a heavy massive cylinder. When it rotates with a constant speed, at that time the energy is stored in the flywheel. During the deceleration period of the rotor, the energy is released from the flywheel and further used to drive the connected motor as a generator [37–39]. The advantages of the system include

a. Excellent cycle stability
b. Longer life
c. Low maintenance
d. Less cost

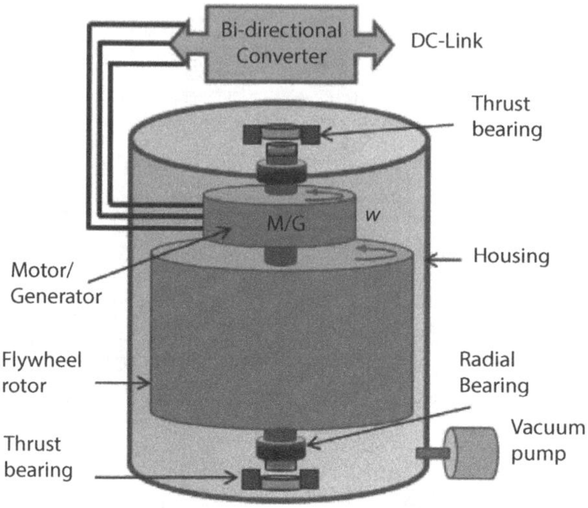

Figure 3.5 Flywheel energy storage [79].

e. Higher power density
f. Higher efficiency

The FES system is not able to supply energy for a longer period.

3.3.4 Chemical Energy Storage

3.3.4.1 Hydrogen-Based ESS

Hydrogen-based ESS or fuel cell (FC) is the advanced storage technology that is free from emission which is shown in Figure 3.6. It converts the

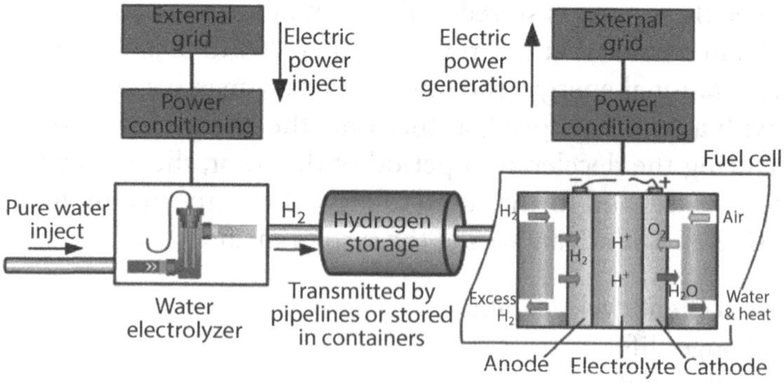

Figure 3.6 Hydrogen-based ESS [78].

chemical form of energy to DC power. It consists of two electrodes which are porous in nature and a layer of electrolyte. The electrolyte may be in the solid-state or in the liquid state. The electrolyte conducts only ion, not the electron. The hydrogen is fed continuously at anode and oxidant at the cathode. Due to the electrochemical reaction, the chemical form of energy is converted to DC power by the electrodes [40]. The main advantage of this system is Pollutant free (water as the only by-product); disadvantages are low efficiency, low power rating, slow response time and difficult to store hydrogen as it is a highly flammable gas [41]. That's why it is used as secondary storage for stand-alone systems [42].

3.3.4.2 Battery Energy Storage (BESS)

In the battery, energy is stored in the electrolyte in the form of electrochemical energy (Figure 3.7). To get the desired voltage and current, batteries are connected in both series and parallel. The cell and battery are made up of anode, cathode, and electrolyte. The electrolyte, which is either liquid, solid, or paste, undergoes a chemical reaction to generate energy. The chemical reaction occurred inside the batteries are reversible in nature. The rate of a chemical reaction inside a battery depends on the load connected to the battery.

As the capacity of an individual cell is less, so for a higher range of application cells are interconnected to form an array. Every cell has a definite rating which depends on the volume/mass of the electrolyte [43, 44].

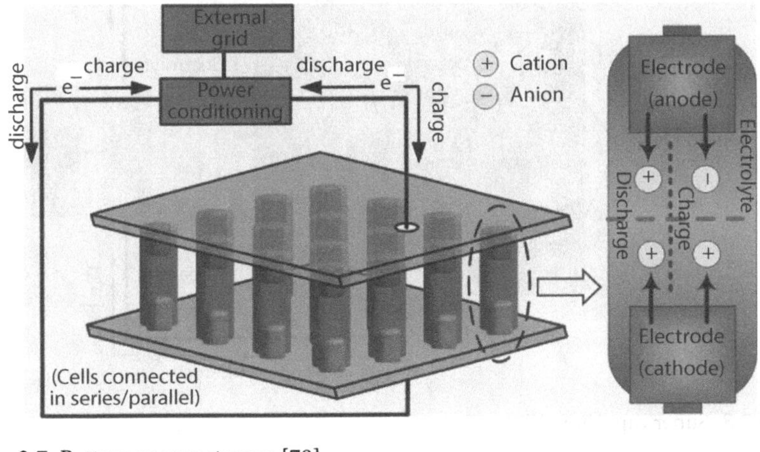

Figure 3.7 Battery energy storage [78].

Batteries are used in stand-alone, grid-connected, and hybrid RES systems. The merit of the battery system is

a. Faster response time
b. Self-discharge loss is small
c. Higher efficiency
d. Energy and power densities is high
e. Technology is mature
f. Available in market in a wide range with varied characteristics

The different types of batteries used in HES system are lead acid batteries, Lithium-based batteries, Nickel-based batteries, Sodium sulfur batteries and redox flow batteries. Every battery system has its own advantages and disadvantages.

3.3.5 Electromagnetic Energy Storage

3.3.5.1 *Super Capacitors (SC)*

The SC is formed by placing two solid conductors, separated by an electrolyte solution as shown in Figure 3.8. The SC capacitance and energy density is much higher than the normal capacitor. The energy discharged/stored within a short period of time is much higher in the case of SC.

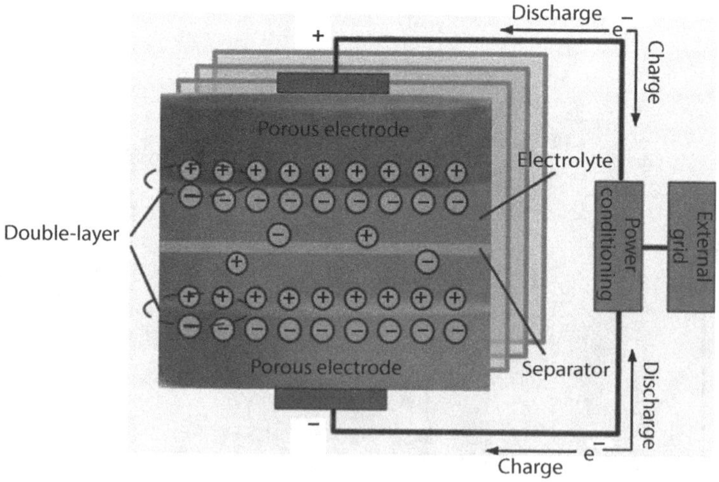

Figure 3.8 Super capacitors [78].

The merits include

a. Cycle stability is unlimited
b. Power density is extremely high
c. Quick rate of charging and discharging
d. Higher reliability
e. Zero maintenance
f. Longer lifetime
g. Wide operating temperature range.

The self-discharge phenomenon is the main draw of SC like flywheels. A super-capacitor is a suitable option for electric vehicle and power quality improvement [45–47].

3.3.5.2 Superconducting Magnet Energy Storage (SMES)

The energy is stored in the magnetic field in SMES as illustrated in Figure 3.9.

The magnetic field is created in the superconductive coil by a DC circulating current which is the main part of SMES. The other parts are vacuum, refrigerator, and power conditioning unit [48, 49]. Until now, SMES was not a fully developed technology.

The advantages of the SMES system are as follows:

a. Faster response
b. Power is available instantaneously
c. Generally utilized for small-capacity storage systems.

The efficiency of different storage technologies is shown in Figure 3.10 and the different characteristic of energy storage technology is shown in Table 3.1.

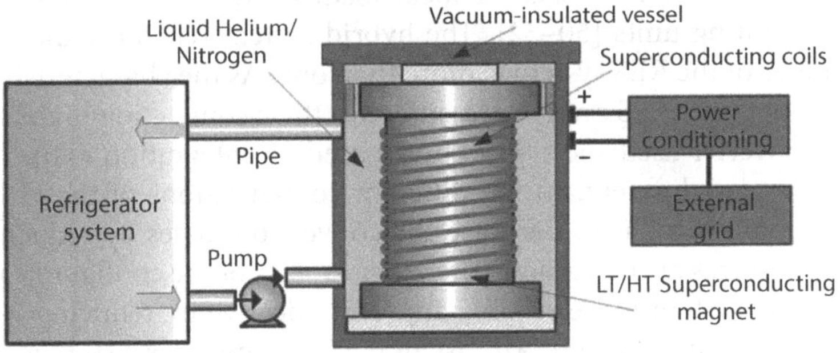

Figure 3.9 Superconducting magnet energy storage [78].

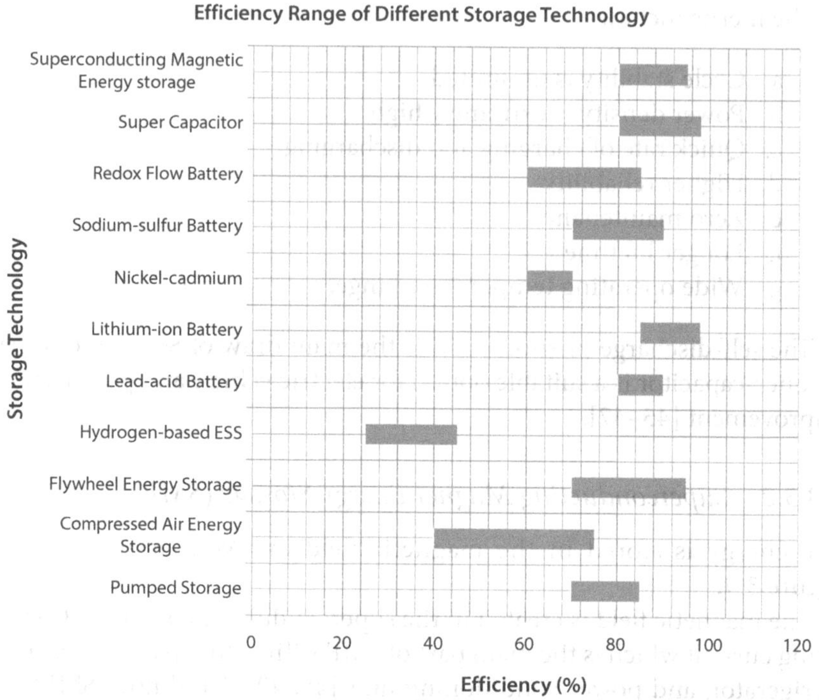

Figure 3.10 Efficiency of different energy storage system.

3.4 Hybrid Energy System Configuration

3.4.1 Integration Schemes

As a renewable energy system is intermittent in nature its characteristic is quite different as compared to the conventional generation system. That's why a well-defined procedure is adopted to connect them in the hybrid configuration which consists of local load, energy storage system, and other generating units [50–52]. The hybrid system reliability increases if the number of the RES system is more than one. As the characteristics and behavior of every RES system are different, their connection architecture, type of converter used and type of buses used for integration to the hybrid system are also different. In the different configurations of the HES system, the main focus is to decrease the conversion losses by reducing the number of stages of conversion. The other aspects of the configurations are to make the system more reliable and cost-effective by reducing its control system complexity [53]. Almost all HES systems have a storage unit to store energy when surplus power is available and can deliver at the time of

Table 3.1 Different characteristic of energy storage technology.

Storage technologies	Power rating (MW)	Cycling	Self discharge (%)	Storage duration	Energy density (Wh/l)	Power density (W/l)	Response time
Pumped Storage	100-1000	30-60 yrs	Approximately zero	4-12 hours	0.2-2	0.1-0.2	Sec-min
Compressed Air Energy Storage	10-1000	20-40Yrs	Approximately zero	2-30 hours	2-6	0.2-0.6	Sec-min
Flywheel Energy Storage	0.001-1	20000-100000	1.3-100	Sec-hours	20-80	5000	10-20ms
Hydrogen-based ESS	0.01-1	5-30 Yrs	0-4	Minutes-week	600	0.2-20	Sec-min
Lead-acid Battery	0.001-100	500–1000	0.1-0.3	1min-8hours	50-80	90-700	<sec
Lithium-ion Battery	0.1-100	1000-10000	0.1-0.3	1min-8hours	200-400	1300-10000	10-20ms
Nickel-cadmium	40	2000–2500	10	sec–min	50-150	210	<20ms
Sodium-sulfur Battery	10-100	2500-4400	0.05-20	1min-8hours	150-300	70-120	10-20ms
Redox Flow Battery	1-100	12000-14000	0.2	2-10 hours	20-70	0.5-2	
Super Capacitor	0.01-1	10000-100000	20-40	ms-min	10-20	40000-120000	10-20ms
Superconducting Magnetic Energy Storage	0.1-1	100000	10-15	ms-min	Approximately 6	1000-4000	<100ms

deficit of power. This can be achieved by connecting the storage unit to the system by a bidirectional converter.

The HES can be operated in two-modes, i.e., in stand-alone mode or grid-connected mode [54]. Different types of architectures of HES are shown in Figure 3.11 [55–58].

3.4.2 DC-Coupled Systems

In this system, the all RES sources are connected to the DC bus with the help of appropriate converters.

In suitable cases, the RES is also connected directly to the DC bus without any converter. In a similar manner, DC loads are also connected to the DC bus either directly or through an appropriate converter according to the requirement. The AC load can be operated from the DC bus by connecting it through an inverter of the appropriate range. A bidirectional inverter can be used to connect it to the grid in case of a grid-connected HES system, which is not shown in the figure.

In this system for integration of different RES, source synchronization is not required, which is why the structure is very simple. The main drawback in the structure, which is shown in Figure 3.12, is that the failure of the inverter leads to a complete shutdown of all AC loads. In order to avoid this, these low-rating inverters are connected in parallel instead of a single large-rating inverter. If multiple inverters are used then synchronization and proper sharing of power is required at the inverter side [59].

3.4.3 AC-Coupled Systems

In this system, all the RES sources are connected to the AC bus with the help of appropriate converters. In suitable cases, the RES is also connected directly to the AC bus without any converter by maintaining voltage and frequency. The AC bus will act as the point of common coupling (PCC)

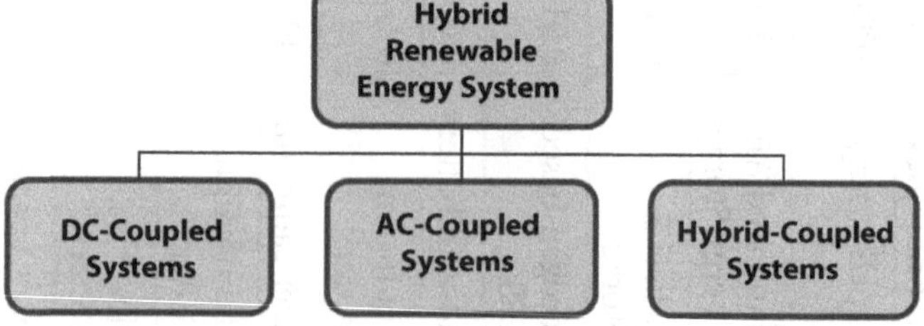

Figure 3.11 Type of HES.

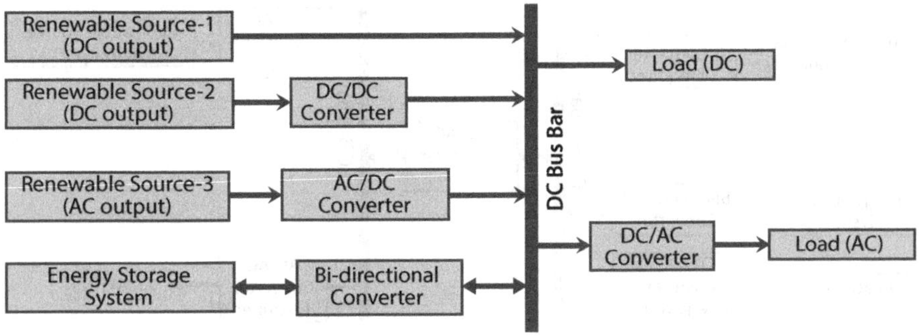

Figure 3.12 DC-coupled systems.

and all the sources have to be synchronized before interconnecting to the PCC [60]. AC loads are connected directly to the AC bus and DC loads are connected through a suitable converter. This type of structure (Figure 3.13) is implemented in the rural area where the AC load demands are more and the system capacity is small [61–63].

3.4.4 Hybrid-Coupled Systems

In hybrid-coupled RES, sources are connected to both AC and DC bus depending on the availability. Further, the AC and DC bus are interconnected through an appropriate bi-directional converter as shown in Figure 3.14. As both AC and DC buses are available, different RES can be connected directly to the system. Due to the direct interconnection of sources, the overall system cost is reduced and its efficiency increases but the control and energy management is more complicated than other coupled configuration. A different coupling system is used according to the availability of the type of RES sources and load.

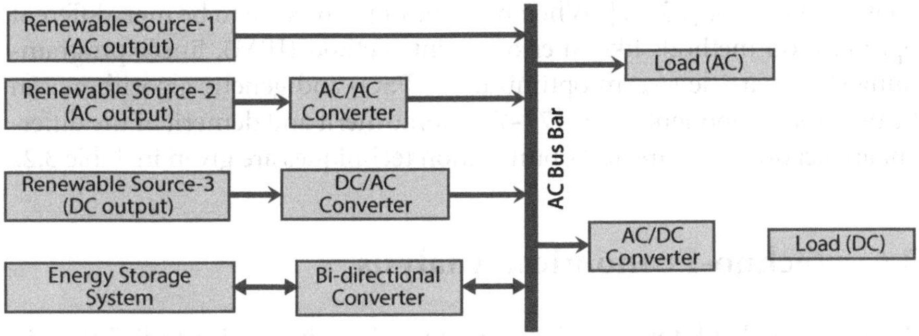

Figure 3.13 DC-coupled systems.

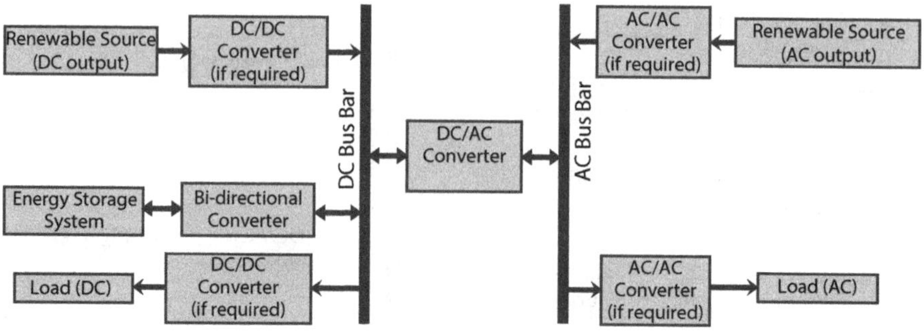

Figure 3.14 Hybrid-coupled systems.

To get more scalability and flexibility in the system the power electronic interfacing circuits shown in Figure 3.7 can be made as modular building blocks.

3.5 Component Sizing of Hybrid RE Systems

In a hybrid energy system, component sizing is an important aspect in order to choose appropriate RES along with suitable ESS for a particular application [64–69]. In most cases, the unit sizing is a complex phenomenon as several constraints and optimization of different objectives are taken into account simultaneously. But in some cases unit sizing can be done by considering the generation technology available and maximum power rating of the equipment. Generally based on the available information of load parameters, generation statistics, required system reliability, geographic factors, cost, and other case-specific information, the unit size can be optimized by considering different objective function like impact of environment, installation and running costs, reliability and payback periods.

Different applications software is available for the proper selection of the above component [70, 71]. When multiple objectives are to be met, different optimization methods like interior- point-method (IPM), linear programming (LP), particle swarm optimization (PSO) and genetic algorithms can be used for component sizing [72–74]. Some merit and demerit of the different application software and optimization techniques are given in Table 3.2.

3.6 Techno-Economical Analysis

A cluster of load for Gahakia remote village in Daringbadi Taluk in Kandhamal district of Odisha has been selected for the study.

Table 3.2 Advantages and disadvantages of some of the optimization techniques.

Sl No	Optimization techniques	Merits	Demerits
1.	PSO	a. Technique is derivative free. b. Not so sensitive to objective function nature. c. Dependency is less. d. Better efficiency rate.	a. Lake of mathematical foundation. b. computational time is larger.
2.	tabu search (TS)	a. Integrated easily with other methods. b. Takes more time in the region where the solution is good. c. The variability in the initial solution is reduced. Due to deterministic moves.	a. Numbers of iterations are more. b. Conversion rate is very slow.
3.	simulated annealing (SA)	a. Non-linear models can be handled. b. Definite optimal solution. c. It can be easily coded for a complex system.	a. Initial constraints are more. b. Final solution is affected by the initial assumption.
4.	harmony search (HS)	a. Easy to implement. b. Parameters adjustment is less. c. Convergence rate is quick.	a. Premature convergence due to complex optimization problems.

(Continued)

Table 3.2 Advantages and disadvantages of some of the optimization techniques. (*Continued*)

Sl No	Optimization techniques	Merits	Demerits
5.	Genetic Algorithm (GA)	a. Multiple solutions exist. b. GA can be easily. implemented in simulations and models.	a. it may stuck in local minima.
6.	power pinch analysis (PoPA)	a. Graphical interface.	a. Only applicable for a single unit.
7.	Electric System Cascade Analysis (ESCA)	a. Multi-generation system can be handled.	a. technique implementation is more complex due to multi-constraint optimization.
8.	Hybrid Optimization Model for Electric Renewable (HOMER)	a. Easier to implement for complex system. b. Graphical representation. make understanding easier. c. cost-effective.	a. Reliability analysis cannot be done.

3.6.1 Selection of Study Area for the Proposed Study

The village is located at $20^0.31'$N and $085^0.47'$ which has a high potential of solar and wind energy. The total geographical area of this village is 60 hectares. Gahakia has a total population of 104 people. According to the 2013 census, this village has about 23 houses. The people of this village depend on agriculture as their major source of income. Although the grid is already connected to this location, the capacity and continuity of electricity supply are limited and the situation in the summer season becomes miserable. Hence, HES is suitable for this village.

3.6.2 Load Assessment of the Study Area

The load demand of the village is calculated based on the needs of the population living in the area. An on-site field survey is conducted and data is collected from the villagers for the different types of loads like domestic load and agricultural load. The major domestic loads are compact fluorescent lamps, fans, TV, fridge water pumps and other home equipments.

The load demand for the agriculture sector comprises four pumps of 5.0 kW each, two pumps of 7.5 KW each and eight grass cutters of 0.75 KW each. Apart from these two types of loads, the deferrable load is also connected in the system to use the surplus energy generated by the HES. The loads which are used for water heating and ice preparing are treated as the deferred loads. The details of the connected load are given in Table 3.3. The power demand by these loads fluctuates throughout the day and year on the basis of the operation of the connected load. Maximum load demand is observed in summer for all sections of the load. A major portion of load demand is mostly in the morning as well as evening time whereas the demand is minimum during night time.

The entire connected domestic load is calculated throughout the year with a load factor of 0.15. Figure 3.15 shows the hourly load profile. The agriculture loads which are also used in daytime have a load factor of 0.25. The maximum load demand for the agriculture sector is observed in the months of June and July because villagers are starting their plantation of different types of crops. Figure 3.16 shows the monthly load demand of the village. Based on these load profiles, the HES are designed accordingly.

3.6.3 Resources Assessment

The potential availability of renewable energy sources like PV and wind energy are taken into account for designing the HES. The availability of

Table 3.3 Estimated load demand of the village.

Sl. No.	Name of load	Power rating (W)	Quantity	Use (h/day)	Energy consumed (Wh/day)
A. Domestic loads					
1	Lamp load	10	80	5	4,000
2	Lamp load	15	40	6	3,600
3	Fan (ceiling)	50	60	16	48,000
4	Fan (table)	75	20	4	6,000
5	Television	150	20	3	9,000
6	Mobile charger	5	40	2.5	500
7	water pump	500	20	1	10,000
8	Refrigerator	150	8	16	19, 200
Total domestic load					100, 300
B. Agricultural load					
10	Water irrigation pump	7500	2	6	90,000
	Water irrigation pump	5000	4	5.5	110,000
	Grass cutting machine	750	8	0.5	3,000
Total agricultural load					203,000
Total load (A+B)					303,300

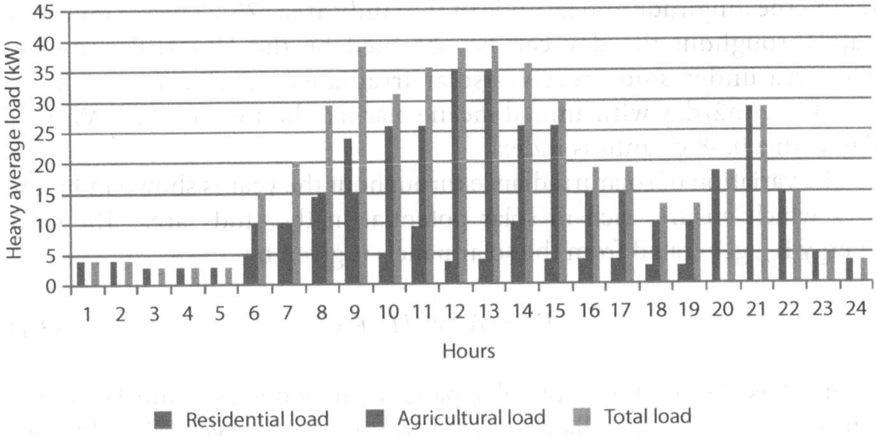

Figure 3.15 The hourly load curve within study area.

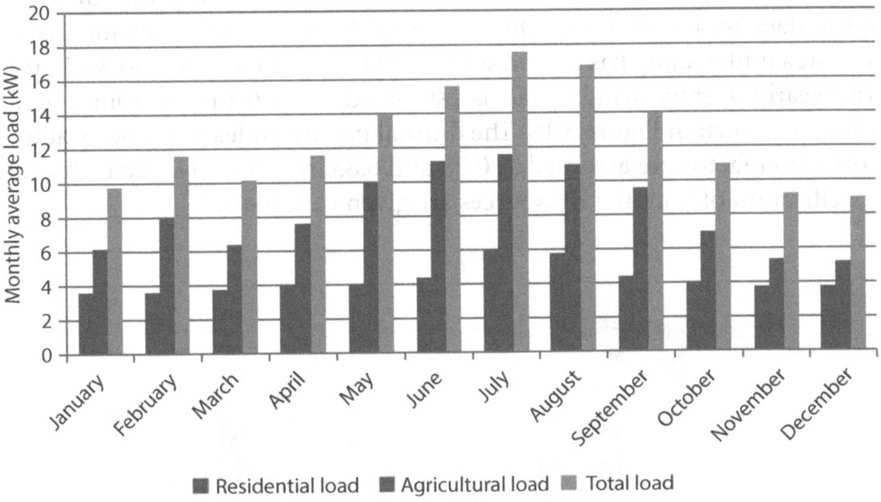

Figure 3.16 The monthly load curve within study area.

solar irradiance is sufficient throughout the year at the village site. However, sunny days during summer have solar irradiance of 11 to 12 hours whereas in winter season it can be accessible for 6-8 hours.

The temperature during summer reaches 46°C in this village. The solar irradiation data used for study at this location was taken from the NASA website (National Aeronautics and Space Administration) by taking the longitude and latitude of the study area [75]. The solar energy data throughout the day can be obtained by the HOMER software. The area under study receives solar irradiance ranging from 5.86 to 9.12 kWh/m2/day with annual mean solar irradiation of 7.12 kWh/m2/day with a 0.58 cleanness index.

The variation of solar irradiance throughout the year is shown in Figure 3.17 which ensures rich in solar potential in the study area. The solar energy E_{PV} generated from the solar panels is given by

$$E_{PV} = A * r * H * PR \tag{3.1}$$

where A is the total area of solar panel in m^2 whereas r and H are the efficiency and solar irradiance in $kWh/m^2/day$ respectively. Here PR represents the performance index of solar panels (0-0.9) of the default level of 0.75.

The average monthly wind speed information for ten years can be collected from the NASA website for the selected location [76]. All wind speed data are taken at a height of 50 m with the help of an anemometer. The area under study has a wind speed in the range of 4.52 m/s to 8.45 m/s. The yearly average wind speed is estimated as 5.66 m/s at 10m height which is shown in Figure 3.18. The diurnal pattern index and hourly autocorrection factor are assumed as 0.25 and 0.85, respectively. The technical specification of both the RE sources are given in Table 3.4.

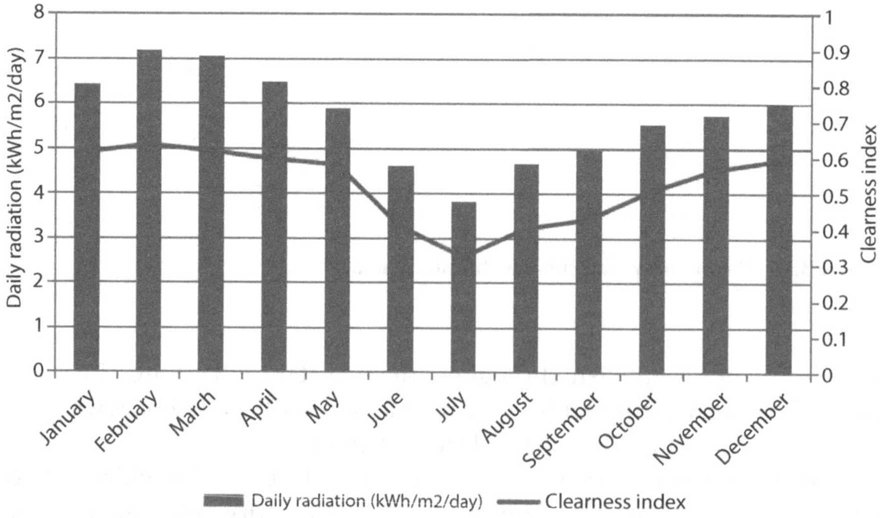

Figure 3.17 Monthly average solar radiation and cleanness index for the study area.

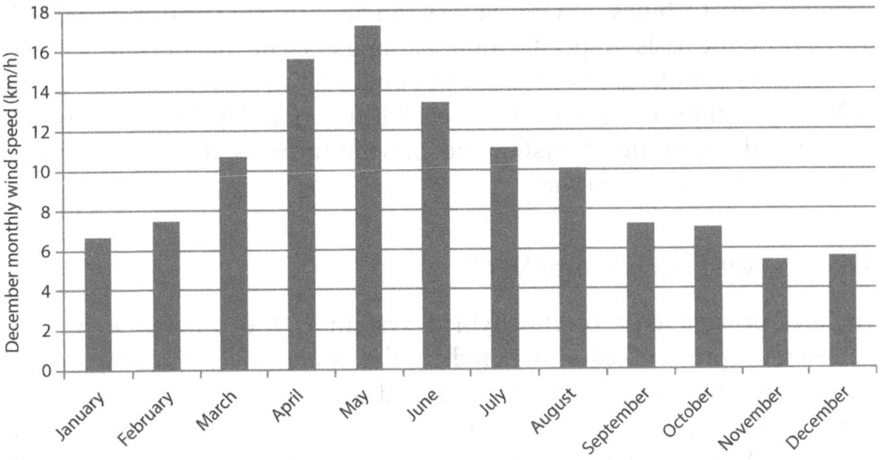

Figure 3.18 Monthly average wind data available.

Table 3.4 Technical components of RE sources.

Technical parameters of the solar PV		**Technical parameters of the wind turbine**	
Parameter	**Value**	**Parameter**	**Value**
Ambient temperature	47 °C	Rated capacity	10 kW
Efficiency	17%	Rated wind speed	12 m/sec
Derating factor	80%	Hub height	24 m
Ground reflection	20%	Cut out speed	25 m/sec
Temperature co-efficient	−0.38%/°C	Cut in speed	3.4 m/sec
Life time	25 years	Capital cost	15000 $/kW
Capital cost	857 $/kW	O & M cost	75 $/year
Replacement cost	857 $/kW	Replacement cost	12000 $/kW
O&M cost	10 $/kW/year	Life time	25 years

3.6.4 Economic Analysis

The realization of HES can be evaluated by specifying various economic parameters like net present cost (NPC) and the cost of energy (COE). The economic and technical aspects of various combinations of solar PV systems, wind turbines, and the battery can be optimized by using HOMER software on the basis of specific input parameters. During the optimization process, the simulation is made with some objective function to reduce COE and total life cycle cost. Based on this function, HOMER determines the optimal size of the PV system, no. of wind turbines, the size of the battery and the converter rating.

3.6.4.1 Net Present Cost (NPC)

The net present cost of the hybrid system is the difference in cost between all present costs incurred by the system over its lifespan and the all-present revenue which is earned during the life period of the project. NPC involves initial capital cost, fuel cost, replacement cost, operation & maintenance (O & M) cost of the HES. Revenue involves all the salvage value.

HOMER computes NPC by applying the following equation:

$$C_{NPC} = \frac{C_{anu,tot}}{CRF(i, R_{proj})} \tag{3.2}$$

Where C_{NPC} and $C_{\text{anu,tot}}$ are the total net present cost (\$) and net annualized cost (\$/year), respectively, whereas CRF represents capital recovery factor, i is the annual rate of interest and R_{proj} (N years) is the life period of the project. The Capital recovery factor (CRF) is given by the following expression:

$$CRF = \frac{i(1+i)^N}{(1+i)^N - 1} \tag{3.3}$$

The nominal interest rate can be expressed in terms of the real interest rate and inflation rate by the expression given below:

$$i = \frac{i_0 - f}{1+f} \tag{3.4}$$

where i, i_0, and f are the real interest rate, nominal interest rate, and annual inflation rate, respectively.

3.6.4.2 Cost of Energy (COE)

The cost of energy (COE) is the average cost per unit (kWh) of useful electricity generated by the HES. As calculated by HOMER, the COE ($/kWh) is the ratio of $C_{ann,tot}$ ($/year) to total useful electricity generated (E_{served}) by the system as per the following equation:

$$COE = \frac{C_{anu,tot}}{E_{served}} \tag{3.5}$$

The HES simulation model is developed in HOMER software by using the input information as specified in Table 3.4. The following assumptions are made during this study:

Annual discount rate = 8%,
Inflation rate = f = 2% and
Life period of the project = N= 25 years

HOMER carries out the simulation for a variety of possible configurations. Three configurations are; i. PV/battery system, ii. Wind/battery system and iii. Wind/PV/battery system.

3.6.5 Results and Discussion

HOMER simulates all three mentioned feasible design configurations on hourly intervals by using the village's load profile and the component parameters. COE and NPC are estimated for the above three cases. HOMER discards the infeasible ones, and ranks the feasible ones on the basis of COE and NPC of the system. Table 3.5 provides details on the specification of each component of HES and their corresponding power generation for all the configurations.

The configuration comprises of Wind/PV/battery system is considered as the optimal configuration among all its counterparts. This configuration has COE of 0.278 ($/kWh) and NPC of $454,242, respectively, as shown in Figure 3.19. The cost of individual components of the hybrid energy system of all three design configurations to meet the same load is represented by the cash flow diagram shown in Figure 3.20. It is confirmed that the wind energy conversion system (WECS) has the highest installation cost and the wind/battery hybrid system is more costly among all three configurations for the same load. From Figure 3.20, it is also confirmed that Wind/PV/battery combination is more economical for the village.

Table 3.5 The hybrid system optimal configurations.

Configuration	PV (kW)	Wind (kW)	No. of battery	Converter (kW)	Generation (kWh/y)	Total load (kWh/y)	Deferred Load (kWh/y)	Unmet energy
PV-Wind-battery	441	20	242	52.1	408816	110016	298800	0
PV-battery	293	0	256	52.8	473464	110016	363448	0
Wind-battery	0	430	654	60.4	432274	110016	322258	0

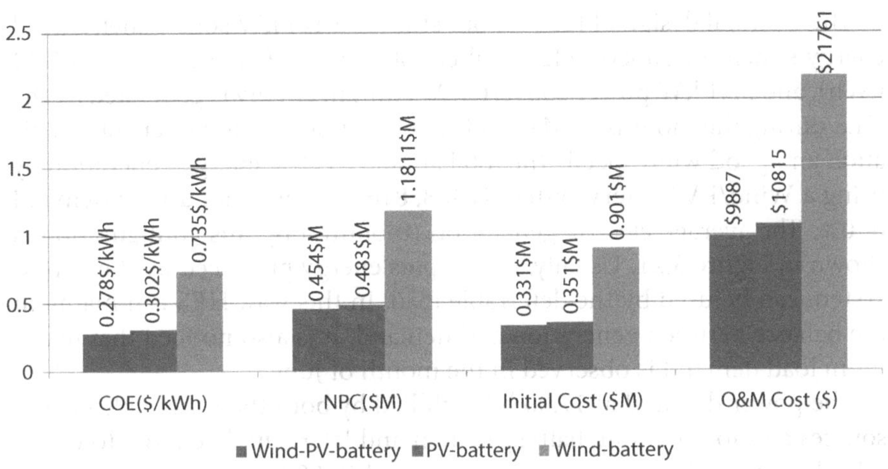

Figure 3.19 Overall cost summary of all three configurations.

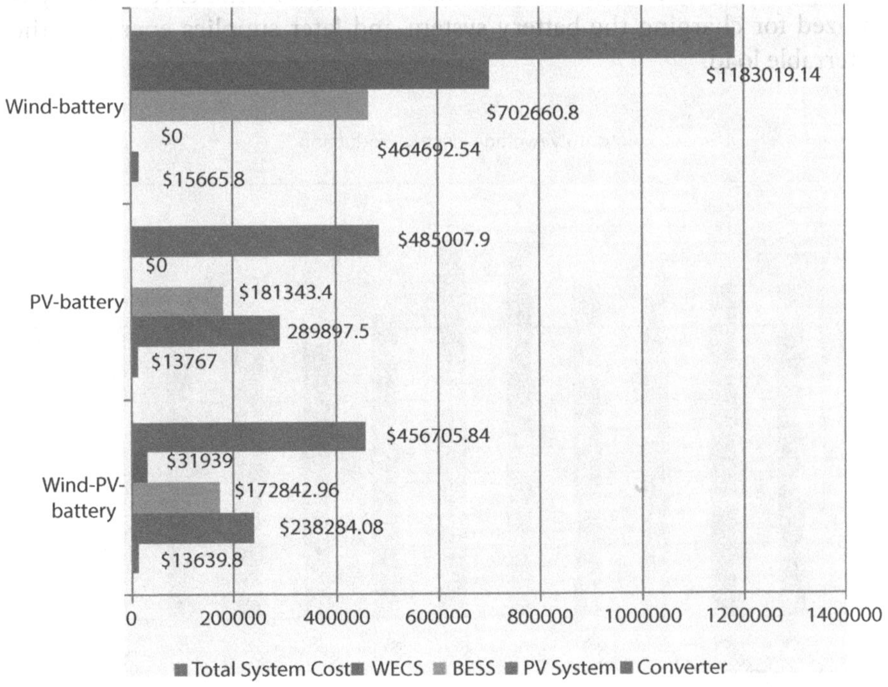

Figure 3.20 Comparison of the economic assessment.

The Optimal design of HES for our study is a 441 kW solar panel, a wind energy system of 20 kW, 242 numbers of batteries (each capacity of 3.12 kWh), and 52.1 kW power converter. Maximum energy is generated by the HES during the months of May and June of a year due to better solar radiation and good wind speed. The total annual electric energy generation by using a Wind/PV/battery system is 408, 816 kWh with an unmet demand of 0%. The average electric generation (monthly) by this configuration is shown in Figure 3.21. Usually, the surplus energy produced by the hybrid system can be used by the deferrable load. In this way, HES can maintain the balance between generation and demand. It is also noticed that maximum load demand is observed in the month of June.

The power demand at 17.00 h is fulfilled by both the renewable energy sources and for charging battery system and later supply energy to deferrable load after charging the battery. Around 18.00 h, the energy generated by the PV system gets diminishing and ultimately reaches zero at 7.00 PM. At this instant, the energy is generated by both wind generator and battery system without supplying any power to the deferrable load. The load demand gets decreasing after 21.00 hr and the energy generated by the wind generator exceeds the load demand at 22.00 h. The excess energy is utilized for charging the battery system and later supplies energy to the deferrable load.

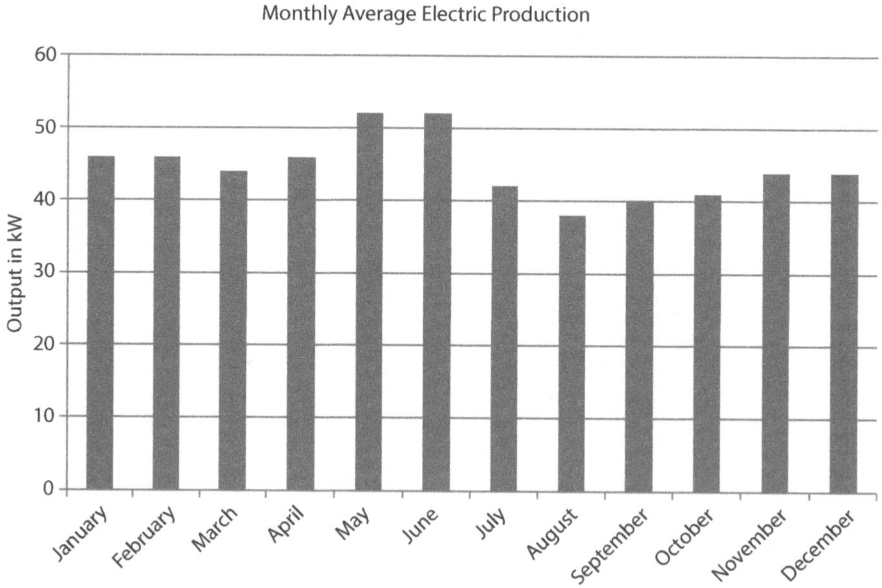

Figure 3.21 Monthly average electricity generation of optimal configuration.

3.7 Conclusion

This chapter focuses on the summary of existing approaches and current research for optimum design of a hybrid renewable energy system for remote locations. Several aspects of design configuration and unit sizing of hybrid energy systems are briefly discussed here. Among the available configuration, ac-coupled systems are more suitable due to fewer power electronic converters and minimum power losses. The reliability of the hybrid energy system can be improved by using the storage units. Different storage units are briefly discussed and their applications are reviewed in this chapter. The techno-economic feasibility is analyzed for a stand-alone hybrid energy system using a case study of a village with the help of HOMER software. Three configurations, namely PV/Battery system, Wind/Battery system and Wind/PV/Battery system, are considered and compared based on the total net present cost and cost of energy in the study area. It can be concluded that the Wind/PV/Battery combination is more economical for the village with COE of 0.278 ($/kWh) and NPC of $454,242.

References

1. Elhadidy, M.A., Shaahid, S.M., 2000. Parametric study of hybrid (wind + solar + diesel) power generating systems. *Renew. Energy* 21 (2), 129–139, 2000.
2. Khare, V., *et al.*, Solar–wind hybrid renewable energy system: A review. *Renew. Sustain. Energy Rev.* 58, 23–33, 2016.
3. Khan, M.J., Yadav, A.K., Mathew, L., Techno economic feasibility analysis of different combinations of PV-Wind-Diesel-Battery hybrid system for telecommunication applications in different cities of Punjab, India. *Renew. Sustain. Energy Rev.* 76, 577–607, 2017.
4. Bhakta, S., Mukherjee, V., Shaw, B., Techno-economic analysis and performance assessment of standalone photovoltaic/wind/hybrid power system in Lakshadweep islands of India. *J. Renew. Sustain. Energy* 7, 2015.
5. A. Al-Sharafi, A.Z. Sahin, T. Ayar, B.S. Yilbas, Techno-economic analysis and optimization of solar and wind energy systems for power generation and hydrogen production in Saudi Arabia, *Renew. Sustain. Energy Rev.* 69 (2017) 33–49,
6. Danish, M.S.S., Senjyu, T., Funabashia, T., Ahmadi, M., Ibrahimi, A.M., Ohta, R., Howlader, H.O.R., Zaheb, H., Sabory, N.R. and Sediqi, M.M., A sustainable microgrid: A sustainability and management-oriented approach. *Energy Procedia*, 159, 160-167, 2019.
7. Fu, Q., Nasiri, A., Solanki, A., Bani-Ahmed, A., Weber, L. and Bhavaraju, V., Microgrids: architectures, controls, protection, and demonstration. *Electric Power Components and Systems,* 43(12), 1453-1465, 2015.

8. Liu, W., Niu, S. and Xu, H., Optimal planning of battery energy storage considering reliability benefit and operation strategy in active distribution system. *Journal of Modern Power Systems and Clean Energy*, 5(2), pp. 177-186, 2017.
9. Bajpai, P. and Dash, V., Hybrid renewable energy systems for power generation in stand-alone applications: A review. *Renewable and Sustainable Energy Reviews*, 16(5), pp. 2926-2939, 2012.
10. Al Busaidi, A.S., Kazem, H.A. and Khan, M.F., A review of optimum sizing techniques for off-grid hybrid PV-wind renewable energy systems. *International Journal of Students Research in Technology & Management*, 2(03), pp. 93-102, 2014.
11. Twidell, J. and Weir, T., *Renewable Energy Resources*. Routledge, 2015.
12. Serrano, E., Rus, G. and Garcia-Martinez, J., Nanotechnology for sustainable energy. *Renewable and Sustainable Energy Reviews*, 13(9), pp. 2373-2384, 2009.
13. Guo, K.W., Green nanotechnology of trends in future energy: A review. *International Journal of Energy Research*, 36(1), pp. 1-17, 2012.
14. Sawin, J.L., Sverrisson, F., Seyboth, K., Adib, R., Murdock, H.E. and Lins, C., *Renewables 2017: Global status report*. REN21, 2017.
15. Zhou, W., Lou, C., Li, Z., Lu, L. and Yang, H., Current status of research on optimum sizing of stand-alone hybrid solar–wind power generation systems. *Applied Energy*, 87(2), pp. 380-389, 2010.
16. Luna-Rubio, R., Trejo-Perea, M., Vargas-Vázquez, D. and Ríos-Moreno, G.J., Optimal sizing of renewable hybrids energy systems: A review of methodologies. *Solar Energy*, 86(4), pp. 1077-1088, 2012.
17. Lupangu, C. and Bansal, R.C., A review of technical issues on the development of solar photovoltaic systems. *Renewable and Sustainable Energy Reviews*, 73, pp. 950-965, 2017.
18. Hau, S.K., Yip, H.L. and Jen, A.K.Y., A review on the development of the inverted polymer solar cell architecture. *Polymer Reviews*, 50(4), pp. 474-510, 2010.
19. De Brito, M.A.G., Galotto, L., Sampaio, L.P., e Melo, G.D.A. and Canesin, C.A., Evaluation of the main MPPT techniques for photovoltaic applications. *IEEE Transactions on Industrial Electronics*, 60(3), pp. 1156-1167, 2012.
20. Maity, S. and Sahu, P.K., Modeling and analysis of a fast and robust module-integrated analog photovoltaic MPP tracker. *IEEE Transactions on Power Electronics*, 31(1), pp. 280-291, 2015.
21. Ashok, S., Optimised model for community-based hybrid energy system. *Renewable Energy*, 32(7), pp. 1155-1164, 2007.
22. Qi, Z. and Lin, E., Integrated power control for small wind power system. *Journal of Power Sources*, 217, pp. 322-328, 2012.
23. Díaz-González, F., Sumper, A., Gomis-Bellmunt, O. and Bianchi, F.D., Energy management of flywheel-based energy storage device for wind power smoothing. *Applied Energy*, 110, pp. 207-219, 2013.

24. Rehman, S., Halawani, T.O. and Mohandes, M., Wind power cost assessment at twenty locations in the Kingdom of Saudi Arabia. *Renewable Energy*, 28(4), pp. 573-583, 2003.
25. Nfaoui, H., Buret, J. and Sayigh, A.A.M., Cost of electricity generated and fuel saving of an optimized wind-diesel electricity supply for village in Tangier-area (Morocco). *Renewable Energy*, 9(1-4), pp. 831-835, 1996.
26. Elhadidy, M.A. and Shaahid, S.M., Role of hybrid (wind+ diesel) power systems in meeting commercial loads. *Renewable Energy*, 29(1), pp. 109-118, 2004.
27. Papadopoulos, D.P. and Dermentzoglou, J.C., Economic viability analysis of planned WEC system installations for electrical power production. *Renewable Energy*, 25(2), pp. 199-217, 2002.
28. Kabir, M.N., Mishra, Y., Bansal, R.C., Coordinated operation of PV and battery energy storage systems, in (Eds.): *Handbook of Clean Energy Systems* (John Wiley & Sons, Ltd, Sussex, UK, 2011.
29. Tan, X., Li, Q., Wang, H., Advances and trends of energy storage technology in microgrid, *Int. J. Electr. Power Energy Syst.*, 44, (1), pp. 179–191, 2013.
30. Kumar, A., Bansal, R.C., A review of multi-criteria decision making (MCDM) towards sustainable renewable energy development, *Renew. Sustain. Energy Rev.*, 69, pp. 596–609, 2017.
31. Adefarati, T., Bansal, R.C., Integration of renewable distributed generators into the distribution system: A review, *IET Renew. Power Gener.*, 10, (7), pp. 873–884, 2016.
32. Ghofrani, M., Arabali, A., Etezadi-Amoli, M., *et al.* Energy storage application for performance enhancement of wind integration, *IEEE Trans. Power Syst.*, 28, (4), pp. 4803–4811, 2013.
33. Koohi-Kamali, S., Tyagi, V.V., Rahim, N.A., *et al.*, Emergence of energy storage technologies as the solution for reliable operation of smart power systems: a review, *Renew. Sustain. Energy Rev.*, 25, pp. 135–165, 2013.
34. Díaz-González, F., Sumper, A., Gomis-Bellmunt, O., *et al.* A review of energy storage technologies for wind power applications, *Renew. Sustain. Energy Rev.*, 16, (4), pp. 2154–2171, 2012.
35. Datta, U., Kalam, A., Shi, J., Hybrid PV–wind renewable energy sources for microgrid application: An overview, In *Woodhead Publishing Series in Energy*, Woodhead Publishing, 2018, Pages 1-22.
36. Chen, H., Cong, T.N., Yang, W., *et al.* Progress in electrical energy storage system: a critical review, *Prog. Nat. Sci.*, 2009, 19, (3), pp. 291–312.
37. Liu, H., Jiang, J. Flywheel energy storage – an upswing technology for energy sustainability, *Energy Build.*, 2007, 39, (5), pp. 599–604.
38. Dufo-López, R., Lujano-Rojas, J.M., Bernal-Agustín, J.L. Comparison of different lead–acid battery lifetime prediction models for use in simulation of stand-alone photovoltaic systems, *Appl. Energy*, 2014, 115, pp. 242–253.
39. Lukic, S.M., Cao, J., Bansal, R.C., *et al.* Energy storage systems for automotive applications, *IEEE Trans. Ind. Electron.*, 2008, 55, (6), pp. 2258– 2267.

40. Barbir, F. *PEM Fuel Cells: Theory and Practice* (Academic Press, Waltham, USA, 2012).
41. Trifkovic, M., Sheikhzadeh, M., Nigim, K., *et al.* Modeling and control of a renewable hybrid energy system with hydrogen storage, *IEEE Trans. Control Syst. Technol.*, 2014, 22, (1), pp. 169–179.
42. Fathima, A.H., Palanisamy, K. Optimization in microgrids with hybrid energy systems – a review, *Renew. Sustain. Energy Rev.*, 2015, 45, pp. 431–446, 2008.
43. Yang, H., *et al.*, Optimal sizing method for stand-alone hybrid solar–wind system with LPSP technology by using genetic algorithm. *Sol. Energy* 82 (4), 354–367.
44. Elhadidy, M.A., Shaahid, S.M., Promoting applications of hybrid power systems in hot regions. *Renew. Energy* 29, 517–528, 2004.
45. Tan, X., Li, Q., Wang, H. Advances and trends of energy storage technology in microgrid, *Int. J. Electr. Power Energy Syst.*, 2013, 44, (1), pp. 179–191.
46. Ghofrani, M., Arabali, A., Etezadi-Amoli, M., *et al.* Energy storage application for performance enhancement of wind integration, *IEEE Trans. Power Syst.*, 2013, 28, (4), pp. 4803–4811.
47. Ahmad Hamidi, S., Ionel, D.M., Nasiri, A. Modeling and management of batteries and ultracapacitors for renewable energy support in electric power systems – an overview, *Electr. Power Compon. Syst.*, 2015, 43, (12), pp. 1434–1452.
48. Y. Sun, Z. Zhao, M. Yang, D. Jia, W. Pei and B. Xu, Overview of energy storage in renewable energy power fluctuation mitigation, in *CSEE Journal of Power and Energy Systems*, vol. 6, no. 1, pp. 160-173, March 2020.
49. Fan, X., Liu, B., Liu, J. *et al.* Battery Technologies for Grid-Level Large-Scale Electrical Energy Storage. *Trans. Tianjin Univ.* 26, 92–103 (2020).
50. H. Nehrir *et al.*, A review of hybrid renewable/alternative energy systems for electric power generation: Configurations, control and applications, *2012 IEEE Power and Energy Society General Meeting, San Diego, CA*, pp. 1-1, 2012.
51. R. Singh and R. C. Bansal, Review of HRESs based on storage options, system architecture and optimization criteria and methodologies, *IET Renewable Power Generation*, vol. 12, no. 7, pp. 747-760, 2018.
52. R. Lasseter, A. Abbas, C. Marnay, J. Stevens, J. Dagle, R. Guttromson, A. S. Meliopoulos, R. Yinger, and J. Eto, *Integration of Distributed Energy Resources: The CERTS Microgrid Concept*, California Energy Commission, P500-03-089F, Oct. 2003.
53. Arul, P.G., Ramachandaramurthy, V.K., Rajkumar, R.K.: Control strategies for a hybrid renewable energy system: A review, *Renew. Sustain. Energy Rev.* 2015, 42, pp. 597–608.
54. Liao, G.-C.: Solve environmental economic dispatch of Smart MicroGrid containing distributed generation system – using chaotic quantum genetic algorithm, *Int. J. Electr. Power Energy Syst.*, 2012, 43, (1), pp. 779–787.

55. K. Agbossou, M. Kolhe, J. Hamelin, and T. K. Bose, Performance of a stand-alone renewable energy system based on energy storage as hydrogen, *IEEE Trans. Energy Convers.*, vol. 19, no. 3, pp. 633–640, Sep. 2004.
56. K. Agbossou, R. Chahine, J. Hamelin, F. Laurencelle, A. Anourar, J.-M. St-Arnaud, and T. K. Bose, Renewable energy systems based on hydrogen for remote applications, *J. Power Sources,* vol. 96, pp. 168–172, 2001.
57. S.-H. Ko, S. R. Lee, H. Dehbonei, and C. V. Nayar, Application of voltage- and current-controlled voltage source inverters for distributed generation systems, *IEEE Trans. Energy Convers.*, vol. 21, no. 3, pp. 782–792, Sep. 2006.
58. F. A. Farret and M. G. Simões, *Integration of Alternative Sources of Energy.* Hoboken, NJ: Wiley, 2006.
59. Sao, C. K. and Lehn, P. W., A transformerless energy storage system based on a cascade multilevel PWM converter with star configuration, *IEEE Trans. Ind. Appl.*, vol. 44, no. 5, pp. 1621–1630, Sep./Oct. 2008, Control and Power Management of Converter Fed Microgrids, *IEEE Trans. Power Systems*, Vol. 23, No. 3, pp. 1088-1098, August 2008.
60. Salas, V., Suponthana, W., Salas, R.A.,Overview of the off-grid photovoltaic diesel batteries systems with AC loads, *Appl. Energy*, 2015, 157, pp. 195–216.
61. Nehrir, M.H., Wang, C., Strunz, K. A review of hybrid renewable/alternative energy systems for electric power generation: configurations, control, and applications, *IEEE Trans. Sustain. Energy*, 2011, 2, (4), pp. 392–403.
62. Scheubel, C., Zipperle, T., Tzscheutschler, P.: Modeling of industrial-scale hybrid renewable energy systems (HRES) – the profitability of decentralized supply for industry, *Renew. Energy*, 2017, 108, pp. 52–63.
63. Diab, F., Lan, H., Ali, S. Novel comparison study between the hybrid renewable energy systems on land and on ship, *Renew. Sustain. Energy Rev.* 2016, 63, pp. 452–463.
64. H. Nehrir *et al.*, A review of hybrid renewable/alternative energy systems for electric power generation: Configurations, control and applications, 2012 *IEEE Power and Energy Society General Meeting, San Diego, CA*, pp. 1-1, 2012.
65. Kellogg, W. D., Nehrir, M. H, G. Venkataramanan, and V. Gerez, Generation unit sizing and cost analysis for stand-alone wind, photovoltaic, and hybrid wind/PV systems, *IEEE Trans. Energy Convers.*, vol. 13, no. 1, pp. 70–75, Mar. 1998.
66. Chedid, R., Akiki, H. and Rahman, S., A decision support technique for the design of hybrid solar-wind power systems, *IEEE Trans. Energy Convers.*, vol. 13, no. 1, pp. 76–83, Mar. 1998.
67. Borowy, B. and Salameh, Z., Methodology for the optimally sizing the combination of a battery bank and PV array in a wind/PV hybrid system, *IEEE Trans. Energy Convers.*, vol. 11, no. 2, pp. 367–375, Jun. 1996.
68. Nelson, D. B., Nehrir, M. H., and Wang, C., Unit sizing and cost analysis of stand-alone hybrid wind/PV/fuel cell power generation systems, *Renew. Energy*, vol. 31, pp. 1641–1656, 2006.

69. Homer, National Renewable Energy Laboratory [Online]. Available: https://analysis.nrel.gov/homer/.
70. DER-CAM—Distributed energy resources customer adoption model, Lawrence Berkeley National Laboratory [Online]. Available: http://der.lbl.gov/dercam.html.
71. Kaldellis, J. K., and Vlachos, G. Th., Optimum sizing of an autonomous wind–diesel hybrid system for various representative wind-potential cases, *Appl. Energy*, vol. 83, pp. 113–132, 2005.
72. Wang, L., and Sigh, C., Multicriteria design of hybrid power generation systems based on modified particle swarm optimization algorithm, *IEEE Trans. Energy Convers.*, vol. 24, no. 1, pp. 163–172, Mar. 2009.
73. Pourmousavi, S. A., Nehrir, M. H., Colson, C. M., and Wang, C., Realtime energy management of a stand-alone hybrid wind-microturbine energy system using particle swarm optimization, *IEEE Trans. Sustain. Energy*, vol. 1, no. 3, pp. 193–201, Oct. 2010.
74. Singh, R., and Bansal, R. C., Review of HRESs based on storage options, system architecture and optimization criteria and methodologies, *IET Renewable Power Generation*, vol. 12, no. 7, pp. 747-760, 2018.
75. NASA, Surface Meteorology and Solar Radiation, 2019. https:// eosweb.larc.nasa.gov/sse/.
76. Tejwani V, Suthar B., Power management in fuel cell based hybrid systems. *Int J Hydrogen Energy*, 42:14980, 2017.
77. Abdellatif, D., AbdelHady, R., Ibrahim, A.M. *et al.* Conditions for economic competitiveness of pumped storage hydroelectric power plants in Egypt. *Renewables* 5, 2 (2018). https://doi.org/10.1186/s40807-018-0048-1 and is used under Creative Commons Attribution 4.0 International License (http://creativecommons.org/licenses/by/4.0/).
78. Xing Luo, Jihong Wang, Mark Dooner, Jonathan Clarke, Overview of current development in electrical energy storage technologies and the application potential in power system operation, *Journal of Applied Energy*, Volume 137, pages. 511 - 536, 2015, ISSN:0306-2619," https://doi.org/10.1016/j.apenergy.2014.09.081,(http://www.sciencedirect.com/science/article/pii/S0306261914010290) and is used under a CC-BY-3.0, (http://creativecommons.org/licenses/by/3.0/).
79. Amiryar, M.E.; Pullen, K.R. A Review of Flywheel Energy Storage System Technologies and Their Applications. *Appl. Sci.* 2017, 7, 286. and is used under a CC-BY-4.0, (http://creativecommons.org/licenses/by/4.0/).

4

Modeling and Energy Optimization of Hybrid Energy Storage System

Hemavathi S.[1,2]

[1]*CSIR - Central Electrochemical Research Institute (CECRI), CSIR-Madras Complex, Tamil Nadu, India*
[2]*Academy of Scientific & Innovative Research, Uttar Pradesh, India*

Abstract
Battery is a main energy storage device that can be widely used in a stand-alone PV system. Due to load transient power and peak load demand, battery lifespan decreases, and design of a large size battery is required. Thus, a high power density characteristic of the ultracapacitor is to be combined with the battery to achieve the high power load requirement. In this article, an energy optimization technique is proposed for a PV hybrid energy storage system which consists of an energy controller unit through which the battery and ultracapacitor are connected in parallel with the same DC bus. The working principle of the energy controller unit is to control and operate the battery and ultracapacitor in suitable modes according to the energy balance of the system which ensures the voltage of DC bus is stable, optimal stable operation of the proposed system and makes the optimal charging and discharging operation of the battery. The proposed energy controller unit effectiveness to be verified in MATLAB simulation and results show the feasibility of the energy optimization strategy.

Keywords: Energy optimization, hybrid energy storage system, li-ion battery, ultracapacitor

4.1 Introduction

The world is moving toward a high volume of oil, and the capability to produce high-quality, cheap, and monetarily extractable oil on demand is

Email: hemavathi@cecri.res.in

Umakanta Sahoo (ed.) Hybrid Renewable Energy Systems, (97–114) © 2021 Scrivener Publishing LLC

decreasing. The ecological effect of non-renewable energy source usage has led to development of sustainable power sources such as solar and wind power [1]. The stand-alone Photovoltaic system is the most common renewable energy source which includes a solar panel, controller, energy storage system, and load [2]. An Energy Storage System (ESS) is utilized to store the surplus power delivered all through the period of high irradiances, or to keep up a steady supply of power to satisfy the load demand during low irradiances in a stand-alone Photovoltaic system [3]. Traditional energy storage systems comprised banks of batteries fit for storing and conveying continuous power to the load [4, 5]. Certain load applications require peak or transient current during starting and regenerative braking and these requirements are supplied from the battery; this degrades its lifetime and increases the size of the battery, which can prove costly. Also, charging of the battery by a Photovoltaic system is unreliable and intensely subject to climate conditions; because of this an optimum charge and discharge cycle cannot be ensured, bringing about a low battery State of Charge (SOC) which reduces the battery life. Therefore, batteries have high energy density characteristics which are suitable to supply stable power [6–8].

To rectify these issues, an alternate energy storage system is combined with batteries capable of providing the peak and burst transient current such as ultracapacitor [9]. By using a battery ultracapacitor hybrid energy system, the battery size can be diminished and a higher SOC can be kept up [10–12]. The ultracapacitor has a higher power density than the battery, which permits the ultracapacitor to give peak power over a short duration of time. On the other hand, the battery has a greater energy density than the ultracapacitor, which permits the battery to store more energy and discharge it over a long duration of time, and extend the battery lifetime by avoiding deep discharge through peak currents [13–15].

The framework proposed in this research article is a Photovoltaic Battery-Ultracapacitor Hybrid Energy Storage System. The energy optimization technique is proposed to control the storage and supply of power all through the system.

4.2 Modeling of Proposed Topology

Hybrid Energy systems consist of different types of energy devices such as energy storage systems, energy generation systems, and renewable energy sources which are integrated into the common DC bus. The proposed PV hybrid energy storage system is shown in Figure 4.1. The PV system is

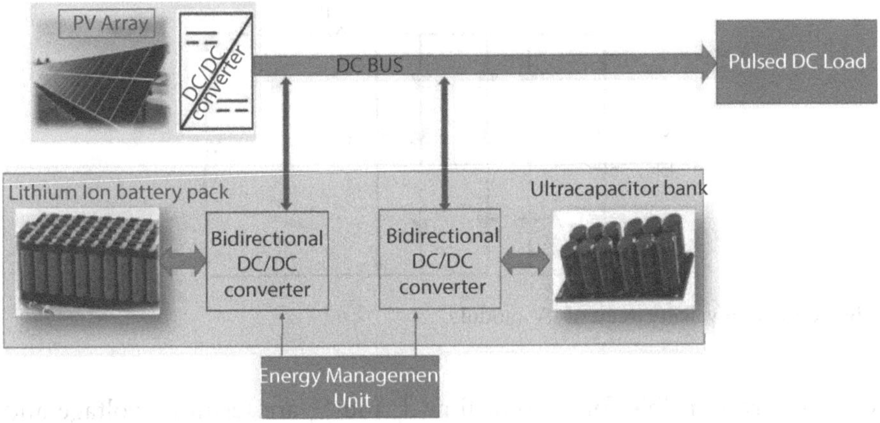

Figure 4.1 Proposed active hybrid energy storage system model.

connected to the DC load (DC grid) through a unidirectional DC-DC converter. Here, the boost converter is designed which tracks the maximum power of the PV system using MPPT technique. A hybrid energy storage system such as battery-ultracapacitor energy storage system is connected in parallel with the same DC bus to DC load through energy management unit. Generally, the hybrid energy storage system is classified based on control strategy such as passive, semi-active, and active energy storage systems. In this topology, the hybrid active energy storage system is proposed which uses separate bi-directional DC-DC converter for battery and ultracapacitor to control the charge and discharge process. So, the proposed system has the capability of fulfilling pulsed load demands based on power distribution between battery and ultracapacitor and ensures the voltage of DC bus is stable. This power distribution between two storage devices can be achieved by the Energy Management unit which controls the bidirectional DC-DC converters located between the storage system and DC load.

4.2.1 Modeling of Photovoltaic System

One of the main components of the hybrid system is PV array which consists of strings of PV modules connected in parallel. Each string consists of modules connected in series. For Mathematical expression, the equivalent circuit of PV module is presented in Figure 4.2.

Where, I_{ph} is photovoltaic Current Source, R_p is parallel resistance which denotes the leakage current, R_s is series resistance with a combination of

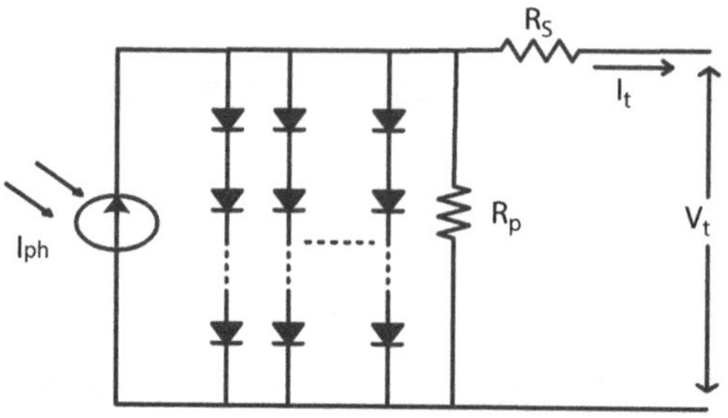

Figure 4.2 Equivalent circuit of PV module.

diodes is responsible for current flow, I_t and V_t are terminal voltage and current of PV module.

The mathematical expression of the equivalent circuit model can be described as follows

$$I_t = n_p I_{ph} - n_p I_0 \left(e^{\frac{q\left(\frac{V_t}{n_s} + \frac{I_t R_s}{n_p}\right)}{AkT}} - 1 \right) - \frac{\frac{n_p V_t}{n_s} + I_t R_s}{R_p} \tag{4.1}$$

The technical specifications of Photovoltaic modules for simulation are selected from datasheet Waaree Energies WS-120 and the modeling parameters are shown in Table 4.1. Based on inputs of PV system such as solar irradiance and cell temperature, I-V and P-V characteristics of one PV module are plotted as shown in Figure 4.3.

4.2.2 Modeling of Li-Ion Battery Module

Li-ion battery is a kind of rechargeable battery that is categorized by different features such as chemistry, performance, cost, and safety characteristics. The equivalent circuit is used to model the Li-ion battery characteristics through circuit elements such as voltage sources, resistors, and capacitors which denote the approximation of electrochemical processes and finding an appropriate mathematical equation that can precisely depict its input to output relation. The equivalent circuit of the single-cell Li-ion battery is shown in Figure 4.4, which is developed based on the Thevenin equivalent model since it has a basic model structure, is flexible, the order number is not high and describes the li-ion battery static and dynamic characteristic

Table 4.1 Modeling specifications of PV module.

S. No	Parameters	Value
1.	Parallel Strings	4
2.	Series connected modules per string	2
3.	Cells per module	72
4.	Maximum Output Power	120 W
5.	Open circuit voltage	21 V
6.	Voltage at maximum power point	17V
7.	Temperature coefficient of open circuit voltage	-0.358%/deg C
8.	Short circuit current	8A
9.	Current at maximum power point	7.1A
10.	Temperature coefficient of short circuit current	0.052A
11.	Solar irradiance	250 – 1000W/m^2
12.	Cell temperature	25°C

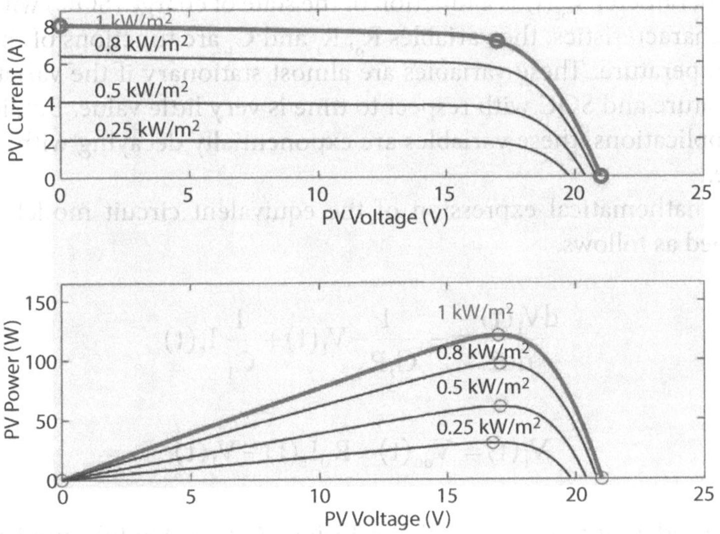

Figure 4.3 IV and PV characteristics of PV module.

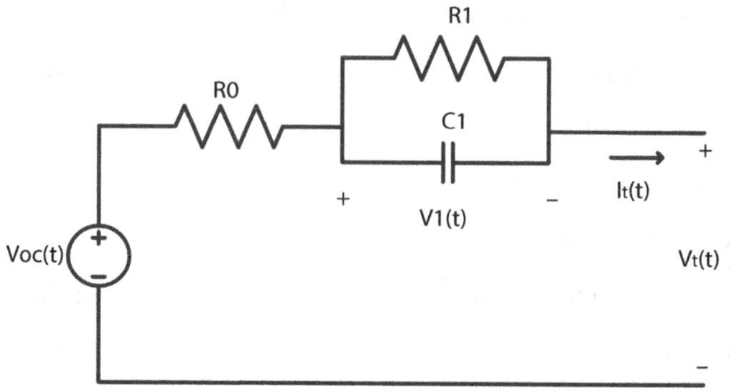

Figure 4.4 Li-ion battery First order Equivalent Circuit Model.

within acceptable modeling errors. Hence, this method has the ability to describe Li-ion battery characteristics in the easiest way.

Where, $V_{oc}(t)$ is open-circuit voltage, R_0 is equivalent internal ohmic resistor, R_1 and C_1 is equivalent polarization resistance and capacitance, which describe the transient response of the Li-ion battery during charging and discharging conditions, $V_1(t)$ is a voltage across the parallel network R_1C_1, $I_t(t)$ is terminal current during charging and discharging, which is considered as the control input parameter, and $V_t(t)$ is a terminal voltage which is considered as the output parameter.

The parameter $V_{oc}(t)$ is a function of the state of charge (SOC) with non-linear characteristics, the variables R_0, R_1 and C_1 are functions of the SOC and temperature. These variables are almost stationary if the variation in temperature and SOC with respect to time is very little value. But, in real-time applications, these variables are exponentially decaying with respect to time.

The mathematical expression of the equivalent circuit model can be described as follows.

$$\frac{dV_1(t)}{dt} = -\frac{1}{C_1R_1}V_1(t) + \frac{1}{C_1}I_t(t) \tag{4.2}$$

$$V_t(t) = V_{oc}(t) - R_0I_t(t) - V_1(t) \tag{4.3}$$

The modeling parameters of the Li-ion Battery Module used in this study are shown in Table 4.2. The Li-ion Battery Module characteristics are protected and monitored by the battery management system.

Table 4.2 Modeling specifications of Li-ion Battery module.

S. No	Parameters	Value
1.	Nominal Pack Voltage	22.2V
2.	Charging Pack Voltage	25.2V
3.	Discharge Cut-off Pack Voltage	18V
4.	Nominal Cell Voltage	3.7V
5.	Charge Cut-off cell Voltage	4.2V
6.	Discharge Cut-off cell Voltage	3V
7.	Nominal Pack Capacity	13Ah
8.	Nominal Cell Capacity	2.6Ah
9.	Standard Discharging Current	6.5A
10.	Initial State of charge	50%

4.2.3 Modeling of Ultracapacitor Module

Ultracapacitor is a kind of double-layer capacitor which has high capacitance value and low voltage limits. The equivalent circuit is used to model the ultracapacitor characteristics through circuit elements such as a series inductor, series resistor parallel resistor and capacitor, which denote the approximation of electrochemical processes and finding an appropriate mathematical equation that can precisely depict its input to output relation. The first order equivalent circuit of the single-cell ultracapacitor is shown in Figure 4.5.

Where, R_s is a series resistance which is also known as equivalent DC series resistance (R_{ESR}) that produces energy loss while charging and

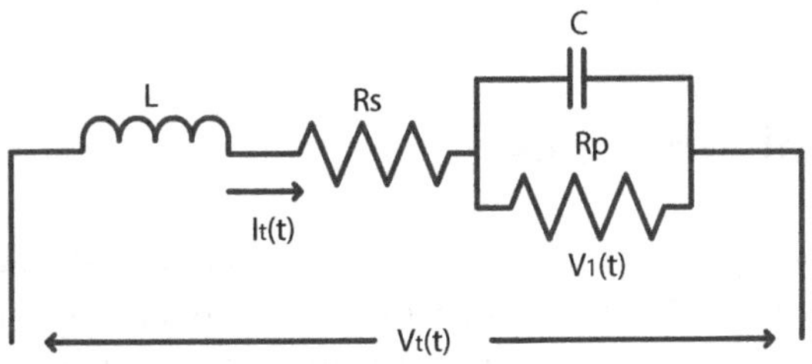

Figure 4.5 Ultracapacitor first order equivalent circuit model.

Table 4.3 Modeling specifications of Ultracapacitor module.

S. No	Parameters	Value
1.	Rated Module Capacitance	29 F
2.	Equivalent DC Series Resistance of Module	0.003Ω
3.	Rated Module Voltage	32V
4.	Rated Cell Voltage	2.7V
5.	Number of cells connected in series	12

discharging the ultracapacitor. R_p is equivalent parallel resistance (R_{EPR}) which is also known as leakage current resistance that produces the energy loss while self-discharge of ultracapacitor and neglected in high-power applications due to higher value than R_s. L is a series inductance that generally denotes the ultracapacitor's physical structure and neglected due to a very small value. $I_t(t)$ is terminal current during charging and discharging, which is considered as the control input parameter, and $V_t(t)$ is a terminal voltage which is considered as the output parameter. C is capacitance of single-cell ultracapacitor, which is composed of a parallel combination of fixed capacitance C_f and variable capacitance C_v dependent on linear voltage ($C_v = kV_1$).

The mathematical expression of the simplified equivalent circuit model can be described as follows.

$$C = C_f + C_v \tag{4.4}$$

$$V_t(t) = R_s I_t(t) + \frac{1}{C}\int I_t(t)dt \tag{4.5}$$

The modeling parameters of Ultracapacitor Module used in simulink model is shown in Table 4.3.

4.3 Control Strategies

The control strategies of proposed PV hybrid energy storage system have two operation modes: MPPT control of PV system and hybrid active power control of Li-ion battery-ultracapacitor energy storage system for control of energy management and stability of the overall system.

4.3.1 PV-MPPT Technique and DC/DC Converter Model

The Maximum Power Point Tracking (MPPT) technique is used to take out the maximum output power from the PV system and the unidirectional DC-DC converter model is designed for transfer of this maximum power to DC grid or batteries for charging based on load demand. The MPPT method of the Perturb and Observe (P&O) algorithm is shown in Figure 4.6.

The algorithm fundamentally gauges the PV voltage and current to compute the maximum power. In the event that the present value of PV power $P_{pv}(t)$ is greater than past $P_{pv}(t-1)$ compared with the present value of PV voltage $V_{pv}(t)$ with the past value $V_{pv}(t-1)$. If $V_{pv}(t)$ is greater than $V_{pv}(t-1)$, the duty cycle is increased one stage or is reduced one stage to extract the maximum power from the PV system. Along with these steps, an optimal duty cycle is gotten to drive the converter to works on Maximum Power Point (MPP).

The Boost DC/DC converter is used in this modeling to step up the voltage and step down the current as per the requirement of DC bus voltage. The circuit diagram of the boost converter and mode of PWM duty cycle operation is shown in Figure 4.7(a) and (b), and the modeling parameters are listed in Table 4.4.

From Figure 4.7(b), The Li-battery State of charge (SOC) and load demand decides the optimum PWM duty cycle of the converter which

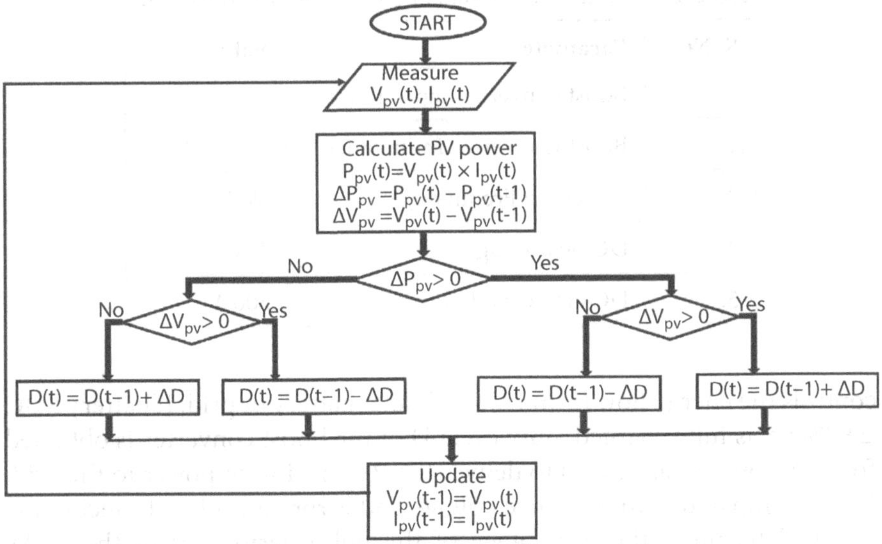

Figure 4.6 Perturb and Observe (P&O) flowchart.

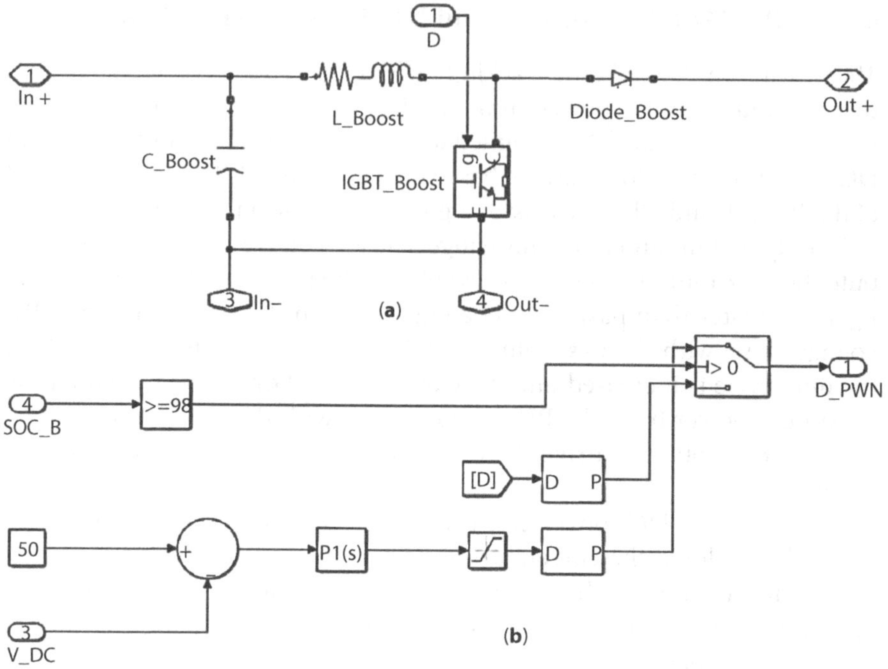

Figure 4.7 (a) Boost Converter Circuit Diagram (b) Selection mode of Duty Cycle to drive boost converter.

Table 4.4 Circuit parameters of DC/DC boost converter.

S. No	Parameters	Value
1.	Boost converter capacitance	2200μF
2.	Boost Converter Inductance	0.352mH
3.	Switching Frequency	5kHz
4.	DC bus voltage	50V
5.	DC active load power	500W

controls the energy flow from PV module to battery or grid. If battery SOC ≥ 98% i.e. is fully charged, duty cycle D of the boost converter is obtained from DC bus voltage (V_{DC}) to deliver the PV maximum power to the grid. V_{DC} is compared with reference voltage and error control by PI technique which determines the percentage of the pulse period (P) to the IGBT switch of a boost DC to DC Converter.

Otherwise, the duty cycle D is obtained from MPPT technique to charge the Li-ion battery, which determines the ON pulse period of the converter.

The capacitor and inductor values of boost converter are described by Eq. 4.4 and 4.5 in continuous conduction mode operation.

$$C_{boost} = \frac{DV_{DC}}{VRf} \tag{4.4}$$

$$L_{boost} = \frac{(1-D)^2 RD}{2f} \tag{4.5}$$

Where, D is PWM duty cycle of boost converter, V_{DC} is DC bus voltage, f is PWM switching frequency, ΔV is the output voltage ripple and R is DC load.

4.3.2 Hybrid Active Power Control of Energy Storage Systems

The active hybrid energy storage system consists of a separate bidirectional DC-DC converter for Li-ion Battery and Ultracapacitor module, and is shown in Figure 4.8. The Buck-Boost DC-DC converter is used in this modeling to control the charge and discharge operation based on power demand.

From Figure 4.8(a) and (b), When Q1 is closed, Q2 is open, the converter acts as a boost converter that discharges the battery and ultracapacitor module and energy flows to the DC bus. When Q1 is open, Q2 is

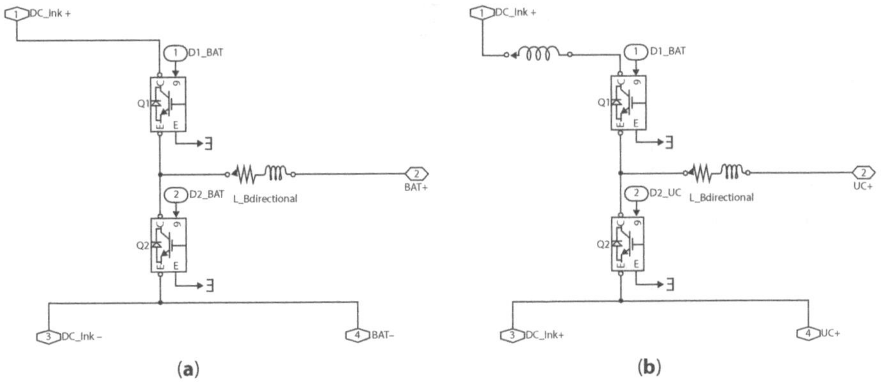

Figure 4.8 Buck-boost DC-DC converter model of (a) Li-ion Battery module (b) Ultracapacitor module.

closed, it acts as a buck converter to control the duty cycle of Q1 to accomplish buck chopper by the anti-parallel diode of the IGBT (Insulated Gate Bipolar Transistor), which charges the battery and ultracapacitor module and energy flows from the DC bus to energy storage system.

The control diagram of buck-boost converter is shown in Figure 4.9. Both buck and boost circuit operation is controlled by double loop control design. The constant voltage control is outward loop control and current-constrained control is inward loop control.

The outward loop control balances the voltage of DC bus (V_{DC}). The reference DC voltage is given voltage (V_{DC_ref}) which is compared with V_{DC}, the outcome is controlled by the PI controller to get a reference current (I_{ref}) to the inward loop control. The inward loop control guarantees the current of the Li-ion battery (I_{BAT}) not to surpass the maximum current of the battery. I_{ref} from the outward control loop is controlled through a current-constrained regulator. In the event that I_{ref} is greater than maximum current, the outcome I_{b_ref} will be changed into the maximum current; otherwise, it is unaltered. The outcome I_{b_ref} is compared with the I_{BAT} and error is controlled by PI technique which obtained the PWM duty cycle (D_{BAT}) to the IGBT switch of a buck-boost converter.

Hence, the control strategy of Li-ion battery utilizes both control loops such as constant voltage and current constrained control to fulfill the constant DC load demand such that the current cannot be excessively enormous. On the other hand, the ultracapacitor is high power density and it can withstand a huge amount of current without any damage. Hence, the control strategy of ultracapacitor utilizes the constant voltage control

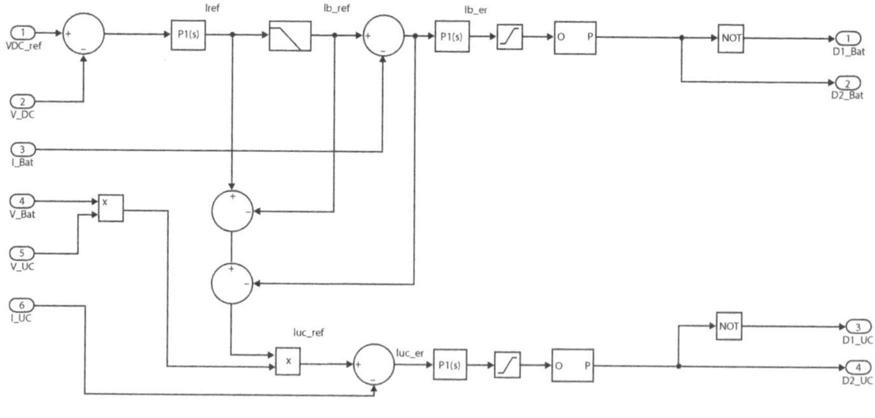

Figure 4.9 Control mode of duty cycle to drive the buck-boost converter of Li-ion battery and Ultracapacitor module.

method only and also ensures the steadiness of the DC bus voltage when the battery is in the current-constrained loop control.

4.4 Energy Optimization Strategy and Simulation Results

4.4.1 Energy Optimization Strategy

The Energy Optimization of Hybrid PV energy storage system is established by operating modes of control strategies such as PV MPPT control and active power energy storage system control.

Because the battery is the most significant energy storage element of the hybrid PV system, the design of the energy controller unit for hybrid PV energy storage system is based on Li-ion Battery switching states. There are four switching states in view of the real-time value of PV power P_{pv}, Load demand P_o and the battery current I_{BAT}, as shown in Table 4.5.

In all the operating states, PV cell operates in MPPT method, which is utilized to consume maximum solar energy.

State I and IV:

When PV system cannot satisfy the load (i.e.) $P_{pv} < P_o$ and battery current $I_{BAT} < I_{BAT\_max}$ (State I), the lack will be supplemented by the discharge

Table 4.5 Switching states of the hybrid PV energy storage system.

Operating conditions	$P_{pv} > P_o$	$P_{pv} < P_o$
$\lvert I_{BAT}\rvert < I_{BAT\_max}$	*State II* PV: MPPT control Li-ion battery: buck constant voltage control Ultracapacitor: buck constant voltage control	*State I* PV: MPPT control Li-ion battery: boost constant voltage control Ultracapacitor: boost constant voltage control
$\lvert I_{BAT}\rvert > I_{BAT\_max}$	*State III* PV: MPPT control Li-ion battery: buck current constrained control Ultracapacitor: buck constant voltage control	*State IV* PV: MPPT control Li-ion battery: boost current constrained control Ultracapacitor: boost constant voltage control

of Li-ion battery and ultracapacitor. In the discharging state, the Li-ion battery and ultracapacitor operate in boost-constant voltage control state which gives the energy from the storage system to DC load and ensures the DC bus voltage stability.

In case $I_{BAT} > I_{BAT\_max}$ (State IV), the excess can be compensated by transforming the Li-ion battery into a boost-current constrained control state to guarantee the battery operating in the desired discharge current.

State II and III:

At the point when the PV system can give more power than the given load power ($P_{pv} > P_o$) due to solar radiation steadily increments and $I_{BAT} < I_{BAT\_max}$ (State II), the excess energy will be complemented by the charge of Li-ion battery and ultracapacitor. In the charging state, the Li-ion battery and ultracapacitor operate in buck-constant voltage control state which gives the energy flow from DC link to the storage system and ensures the DC bus voltage stability.

At the point when the PV power is much greater than given load power and $I_{BAT} > I_{BAT\_max}$ (State III), the excess can be supplemented by transforming the Li-ion battery switching state into buck-current constrained control state for overcurrent protection and ultracapacitor remains operating in buck-constant voltage control state because it could withstand high current without any harm.

4.4.2 Simulation Results

The simulation model of proposed energy optimization control strategy of hybrid PV energy storage system is designed in MATLAB according to the topology as shown in Figure 4.1 and design specifications in Tables 4.1–4.4. The reliability and effectiveness of the proposed strategy is to be verified using simulation results as shown in Figure 4.10.

When the proposed topology is operated in different switching states, the simulation results of voltage, current, and power waveforms of PV system, DC load, and Hybrid Li-ion battery-Ultracapacitor energy storage system is shown in Figure 4.10(a), (b) and (c). For better clearance, the charging current of Li-ion battery is indicated as a negative value as well as the charging current of ultracapacitor is indicated as a positive value. Also, Figure 4.10(d) shows the state of charge (SOC) level of Li-ion battery and ultracapacitor module in different modes of operation. In the simulation model, the initial level of SOC taken for Li-ion battery is 50% and ultracapacitor is fully charged.

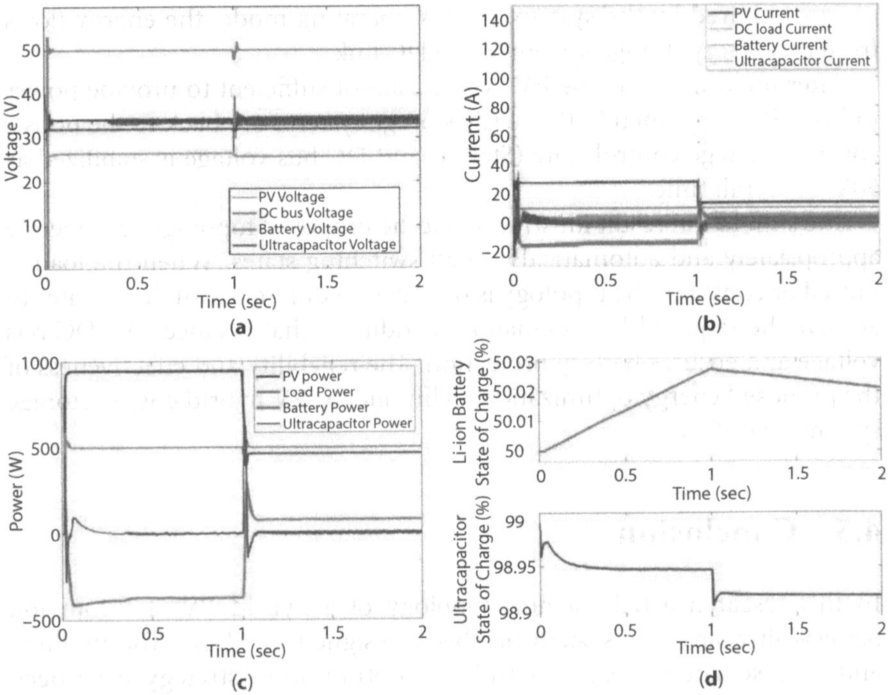

Figure 4.10 Simulation Results of PV Hybrid Energy Storage Topology in different operating modes (a) Voltage Waveforms, (b) Current Waveforms (c) Power Waveforms (d) SOC Waveforms of Li-ion Battery and Ultracapacitor module.

Before the time 1 sec, the PV system can give more power than required load power 500W ($P_{pv} > P_o$), then the excess energy is complemented by the charge of the Li-ion battery (- ve) and ultracapacitor (+ ve). In the ultracapacitor module, the initial SOC is nearly 99% which indicates it is about fully charged and slight jerk happened due to load transient and it reaches its steady-state without any harm and maintains the stable DC bus voltage of 50V as shown in Figure 4.10(a). Thus, Li-ion battery works in buck-constant voltage control state (State II) and the energy flows from DC bus to energy storage system.

At the time of 1 sec, the PV power is gradually decreased, then the lack of energy is supplemented by the discharge of the Li-ion battery (+ ve) and ultracapacitor (- ve). When the battery current reaches its maximum value, the battery transforms into a boost-current constrained control state for protecting the battery from overcurrent and ultracapacitor works in boost-constant voltage control state (State IV) to satisfy the remaining energy required by the system. In this operating mode, the energy flows from the energy storage system to the DC link.

After the time 1 sec, the PV system cannot sufficient to provide power to load ($P_{pv} < P_o$), then both energy storage systems are back to the boost-constant voltage control state (State I) and DC bus voltage is stabilized as 50V in overall time.

The simulation result illustrates that the proposed topology can operate appropriately and automatically in all switching states. When the load is pulsed or constant, the topology is operated in different switching states to achieve the required load demand; in addition, that balances the DC bus voltage and ensures battery protection. The reliability and effectiveness of the proposed energy optimization technique for PV hybrid energy storage system is verified.

4.5 Conclusion

In this research article, a new topology of a hybrid PV- Lithium ion battery/ultracapacitor system has been designed in MATLAB/Simulink, and its associated energy control and optimization strategy have been actively developed for enhancing the life cycle of Li-ion battery. The proposed strategy has been tested in a PV hybrid energy storage system under various switching conditions. The simulation results show the effectiveness and reliability of the proposed energy controller unit and the benefits of using the active hybrid energy storage system under different loading conditions. The results confirm that (i) Li-ion battery is associated with a bidirectional DC/DC converter, which controls the battery over-discharge current and increases the battery lifetime by constant voltage and current constrained control method, (ii) ultracapacitor is associated with a bidirectional DC/DC converter which satisfies the load peak/transient current by constant voltage control method, (iii) PV system is associated with MPPT control, which optimizes the use of solar energy, and (iv) based on energy management strategy, the proposed topology has been operated in optimal energy balance and ensures the voltage of DC bus is stable.

Acknowledgment

The author would like to thank the Research and development in battery division of Central Electrochemical Research Institute, Council of Scientific and Industrial Research in India, for financial assistance. The work presented in this paper is a part of the In House Project (IHP) "Design and Development of Battery Management System for Li-ion Battery Pack for Hybrid Energy Storage System", funded by Central Electrochemical Research Institute, India.

References

1. U. M. Choi, K. B. Lee and F. Blaabjerg, "Power electronics for renewable energy systems: Wind turbine and Photovoltaic systems", *International Conference on Renewable Energy Research and Applications*, Nov. 2012.
2. Salman Salman, Xin AI and Zhouyang WU, "Design of a P-&-O algorithm based MPPT charge controller for a stand-alone 200W PV system", *Protection and Control of Modern Power Systems* 3, 25 (2018).
3. S.M. Ali, Punyashree Pattanayak, Prasan Sanki and R.R. Sabat, "Modeling and control of grid connected hybrid photovoltaic distributed generation system", *International Journal of Engineering Sciences and Emerging Technologies*, Vol. 4, No. 1, pp.125-132, Dec. 2012.
4. Koki Ogura, Mohan Lal Kolhe, "Battery technologies for electric vehicles", *Electric Vehicles: Prospects and Challenges*, pp. 139-167, 2017.
5. Xiaoqiang Zhang, Weiping Zhang, Geyang Lei, "A Review of Li-ion Battery Equivalent Circuit Models", *Transactions on Electrical and Electronic Materials*, Vol. 17, No. 6, pp. 311-316, December 25, 2016.
6. Zhengming John Zhang, Premanand Ramadass, Weifeng Fang, "Safety of Lithium-Ion Batteries", *Lithium-Ion Batteries Advances and Applications*, pp. 409-435, 2014.
7. Yagmur Kircicek, Ahmet Aktas, Mehmet Ucar, Sule Ozdemir and Engin Ozdemir, "Modeling and analysis of a battery energy storage system supplied from Photovoltaic power source", *7th International Ege Energy symposium and Exhibition, June 2014.*
8. Hemavathi S, Manohar Singh, "Microgrid Short Circuit Studies", *IEEE 8th Power India International Conference (PIICON), Dec 10-12, 2018.*
9. A. B. Cultura II, Z. M. Salameh, "Modeling, Evaluation and Simulation of a Supercapacitor Module for Energy Storage Application", *International Conference on Computer Information Systems and Industrial Applications*, pp. 876-882, 2015.
10. M.E. Glavin, Paul K.W. Chan, S. Armstrong, and W.G Hurley, "A Stand-alone Photovoltaic Supercapacitor Battery Hybrid Energy Storage System", *13th International Power Electronics and Motion Control Conference, Sep. 1-3, 2008.*
11. Pranoy Kumar Singha Roy, H. Bora Karayaka, Yanjun Yan, Yazan Alqudah, "Investigations into best cost battery-supercapacitor hybrid energy storage system for a utility scale PV array", *Journal of Energy Storage* 22, pp. 50-59, 2019.

12. Vrinda Tibude, Sneha Tibude, "Hybrid Energy Storage System consisting Solar Panel – Battery Super capacitor for improving the performance of Electric Vehicles", International *Journal of Innovative Science, Engineering & Technology*, Vol. 3 Issue 1, pp. 217-221, January 2016.
13. Wenlong Jing, Chean Hung Lai, Wallace S.H. Wong, M.L. Dennis Wong, "Dynamic power allocation of battery-supercapacitor hybrid energy storage for standalone PV microgrid applications", *Sustainable Energy Technologies and Assessments* 22, pp. 55-64, 2017.
14. Yu Zhang, Zhenhua Jiang, "Dynamic Power Sharing Strategy for Active Hybrid Energy Storage Systems", *IEEE Vehicle Power and Propulsion Conference*, Sep 7-10, 2009.
15. Sathish Kumar Kollimalla, Mahesh Kumar Mishra, Lakshmi Narasamma N, "A New Control Strategy for Interfacing Battery Supercapacitor Storage Systems for PV System", IEEE Students' *Conference on Electrical, Electronics and Computer Science*, 2014.

5

Techno Commercial Study of Hybrid Systems for the Agriculture Farm Using Homer Software

Sanjay Kumar C*, Karthikeyan M, Prasannakumaran K M and V. Kirubakaran†

Centre for Rural Energy, The Gandhigram Rural Institute - Deemed to be University, Gandhigram, Tamilnadu, India

Abstract

There is always an inter-linkage between Agriculture, Water and Electricity. The Energy requirement in the Agriculture sector has increased from 16092 × 10^6 Mega Joules late 1970's to 748047.6 × 10^6 Mega Joules in March 2019 and the demand is still increasing in the future. This paper analyses the possibility of using Hybrid systems for producing low-cost electricity to satisfy the needs of the Agriculture farm. Two different types of hybrid systems are proposed (Stand-alone PV with Gasifier or Stand-alone PV with Biogas) to satisfy the electricity demand of a Farm. Depending upon the Farm, the proposed model can be implemented. Design, Sizing, Optimizing based on the Area and Demand are very important critical parameters in Hybrid Renewable Energy System (HRES) and are carried out in this paper. The HOMER (Hybrid Optimization Model for Electric Renewable) software had been used as a simulation tool for a Hybrid system. This estimates the Techno-Economic Analysis, Installation and operating cost of this proposed system.

Keywords: PV system, biogas plant, biomass, design, HOMER software

**Corresponding author*: csksanjay10@gmail.com
†*Corresponding author*: kirbakaran@yahoo.com

Umakanta Sahoo (ed.) Hybrid Renewable Energy Systems, (115–134) © 2021 Scrivener Publishing LLC

5.1 Introduction

Usage and Demand for fossil fuel are increasing and have led to pollution, GHG emission and climate change, and changes in our day-to-day life. These changes not only affect the environment but also highly impact the Agriculture sector. The electricity demand requirement of the agricultural sector is one-third of the total electricity demand of the country and it greatly impacts economic growth. To reduce the usage of conventional energy such as fossil fuel for the production of electricity in the agricultural sector, the government of India has a target of 175MW of total installed capacity of Renewable energy technologies by the end of 2022. There is a need to develop the Renewable energy technology to meet the electricity demand in the respective agricultural farms. This leads to net zero carbon emission, which helps to attain the sustainable environment. In the agricultural sector efficient utilization of organic waste at the earliest stage of waste generation is far greater than managing (collection, separation and converting) the agricultural waste at the final stage or disposing of the waste as landfill/open dumping or open burning. Due to weather conditions, the instability of power production in PV (Photovoltaic panels) alone cannot meet the total demand of the particular farm and effective utilization of waste material in the farm to produce the electricity. The best option is hybrid Renewable Energy technology case 1: Integration of Stand-alone PV with Biomass (Gasifier) in case dry waste is available, and case 2: Integration of Stand-alone PV with Biogas plant in case wet waste is available.

In India, the electricity requirement for the agricultural sector is given in subsidy rates with suitable supply times which is less than 10 hours a day, which is inadequate and unreliable. Many surveys show that agricultural farms receive fewer hours of electricity than the scheduled time for several purpose with high fluctuation and high interception which leads to the pumps' burn out and need of repair. Farmers had made expenditures to resolve the problem in the form of huge investment in diesel pump sets or electrical pumps. Poor farmers are highly affected by the power quality problems and higher tariff rate by the DISCOM.

In this paper solar is the primary energy source and when the peak load exceeds 1.2 times the additional energy is utilized by converting the biomass into energy. By considering the facts like topography, land usage, natural resources, type of crop cultivated and climatic condition the biomass conversion technology varies (Biogas plant-starch content crops/biomass, Gasification-dry biomass). Initially the necessary basic results are analyzed using the HOMER software. Simulation, Optimization, and Sensitivity Results (automatically run it could be additionally optimized using input values, that are 50% higher and lower than the base case for certain variables) are being processed and Refined the optimized results were found by using HOMER's Search Space. Resource Assessment for this proposed system has been included in this paper for reducing life cycle cost and considering the environmental impact.

5.2 Electricity Consumption by Agricultural Sector

This diagram shows that the electricity consumption of agricultural sector over the period of 8 years is plotted against the growth of the country. As per records of CEA (Central Electricity Authority) a total of 21.3 million pump sets were energized in the year 2018-2019.

5.3 Literature Review

Hossam *et al.* (2019) [8]: This paper deals with the integrating of Grid-Tied PV Biomass Hybrid Renewable Energy System for a small building in Egypt. Comparison is carried out by simulation in HOMER software between the conventional energy vs. Hybrid system. It shows that electricity distribution of 69% of Biomass, 25% from PV and 6% from grid is the best option to supply the electricity, which shows that the grid becomes less important than other power sources. The building is continuously supplied with the required load with suitable back power option.

J.G. Castellanos *et al.* (2015) [11]: This paper investigates the electrical load requirement up to 22 MWh $year^{-1}$ which is satisfied by the off-grid integrated Renewable Energy system for Rural electrification of a village in the West Bengal region, India. Techno-economic performance for different scenarios is estimated in HOMER software with daily variation in the demand, and comparison is based on the Cost of Energy (COE).

S.Baharam *et al.* (2016) [10]: Several tools are available for economical and feasibility to analyze the hybrid. This paper reviews the best option for Hybrid Renewable energy system (HRES) which is well optimized by the HOMER software. The key points to mention include load range from 0.6KW to 2,213,000 KW, and a wide range of combination dispatchable/Non-dispatchable resources, storages and converters have been modeled.

5.4 Study Location

Figure 5.2 shows the how much solar radiation received at particular location with an longitude of 77.89 and latitude of 10.43 (Dindigul District). The Global Horizontal irradiance and Direct normal irradiance data is sampled at an interval of one second and averaged at an interval of one minute. Table 5.1 shows the annual solar radiation profile and Figure 5.2 shows the graph which is plotted against year, Clearness index and Daily solar radiation and Monthly solar radiation profile of Dindigul district. Table 5.2 shows load requirement of the particular farm on the daily basis. Dindigul has a major area of cultivation of grape of about 191.ha which has a share of 8% of the total production of the crop [1]. For this study a grape farm which is located in the Dindigul district nearby sirumalai is selected. The reason for choosing the fruit crop which has huge potential of foreign exchange but in the last 10 years area of cultivation and productivity has much declined is due to several reasons. To increase the cultivation area and profit of the farmers, and depending upon the sources available at the site, Hybrid Renewable Energy Technology, which reduces the financial burden of the farmers, should be implemented. Depending upon the land pattern the farm comes under the medium farm (3 ha) which includes 10 cows.

5.4.1 Solar Energy Potential in Dindigul District

Figure 5.1 shows that the electricity consumption of agricultural sector over the period of 8 years is plotted against the growth of the country. As per records of CEA (Central Electricity Authority) a total of 21.3 million pump sets were energized in the year 2018-2019.

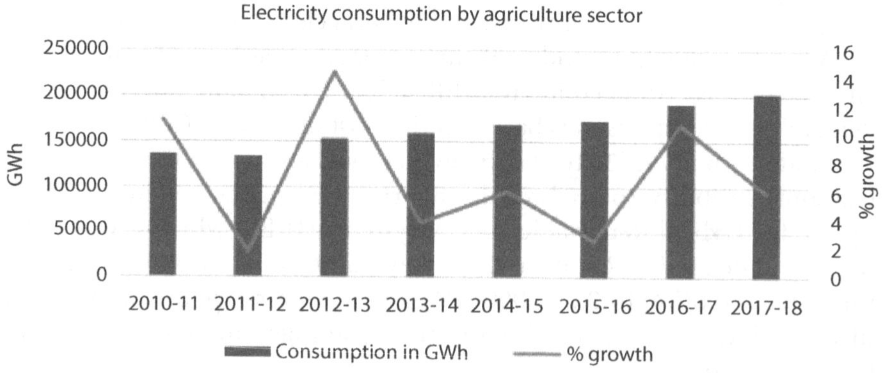

Figure 5.1 Energy consumption in agricultural sector in India.

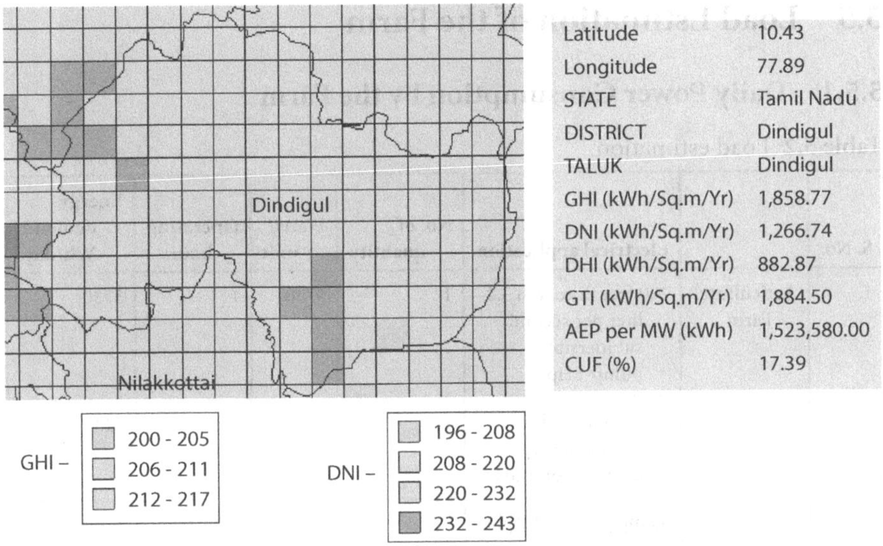

Figure 5.2 Solar radiation distribution in Dindigul District. Source: Data obtained from National Institute of Wind Energy [13].

Table 5.1 Monthly average solar radiation and temperature profile for Dindigul District.

Month	Solar radiance kWh/m²/day	Clearness index	Temperature °C
January	5.11	0.57	23.4
February	6.17	0.64	25.5
March	6.48	0.63	27.3
April	6.32	0.54	26.2
May	5.65	0.47	25.6
June	4.91	0.47	24.7
July	4.94	0.48	24.5
August	5.02	0.52	24.6
September	5.35	0.52	25.0
October	4.83	0.50	24.4
November	4.67	0.52	23.6
December	4.64	0.54	23.1

5.5 Load Estimation of the Farm

5.5.1 Daily Power Consumption by the Farm

Table 5.2 Load estimation.

S. No.		Electrical application	No. of quantity	Watts/ unit	Operating hours	Energy demand Wh/day
1.	Agricultural Farm	700 feet Bore well 3.2 liter per second submersible pump-5Hp	1	3730	1	3730
		500 ft Bore well residential type compressor-1.5Hp	1	1119	2	2238
		Lamp post (Led Light)	5	40	6	1200
		Total (1) = 7168				
2.	Farm House	0.5 hp domestic water	1	373	2	746
		Tube light	2	55	5	825
		Table Fan	2	25	4	200
		Led Light	3	40	5	600
		Total (2) = 2371				
3.	Storage House	Led light	3	55	4	660
		Tube light	1	30	4	120
		Total (3) = 780				
Total Demand of the farm = 10319Wh/day = 10 KW						

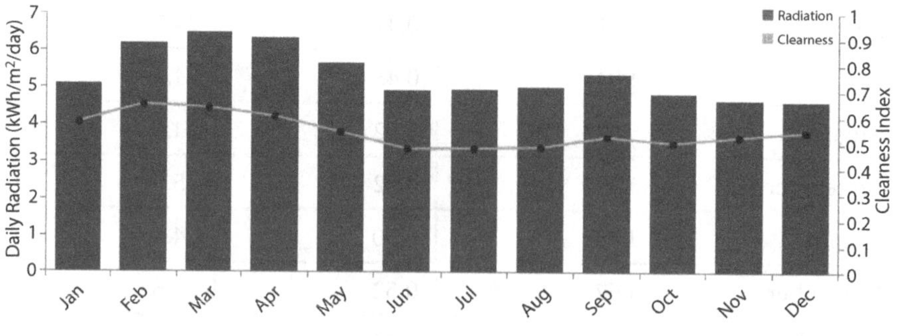

Figure 5.3 Solar radiation profile: Source: National Renewable Energy Lab database [12].

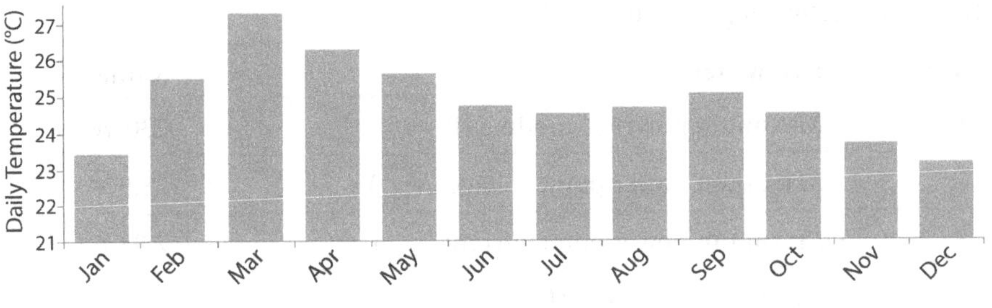

Figure 5.4 Monthly average temperature distribution in Dindigul District.

5.6 Renewable Energy Technology Used in the Hybrid System

5.6.1 Solar PV System

Solar Photovoltaic system is used to convert solar radiation directly into electricity for the utility purpose which can be an on-grid or off-grid system. To determine the efficient working of the PV system, the correct design and sizing of the components is essential. Daily load requirement, Total demand and hours of sunshine are to be considered while designing the PV system. This paper describes two different hybrid system which use energy conversion system such as the Solar Photovoltaic system, Bio-mass gasification system, Bio-gas digester gas engine to produce the electricity. How to determine the sizing of technologies is shown below.

5.6.1.1 PV Module

Two different studies were carried out. Case 1: Solar module of 5 KW has been hybrid with gasifier of 8KW. Case 2: solar module of 6 KW has been hybrid with the 12m^3 of biogas plant to meet the requirement of the farm. PV module is Mono crystalline type and its technical parameters is shown below. Table 5.3 shows the technical parameters of Photovoltaic module.

5.6.1.2 Storage Batteries

To store excess current from the PV when the solar isolation is high during the daytime, Batteries play a vital role in the PV system. HOMER software indicates the type of battery which has been used in the system.

Table 5.3 Technical parameters of PV module.

S. No.	Parameter	Value
1.	Maximum power of module (P_{max})	230 W
2.	Maximum Power point voltage (V_{mpp})	28.5 V
3.	Maximum power point current (I_{mpp})	7.93 A
4.	Short circuit current (I_{sc})	8.52 A
5.	Open Circuit Voltage (V_{oc})	36.8 V
6.	Maximum system voltage	1000 V

While sizing the battery days of autonomy is 3 days and Nominal voltage of the battery is considered 12V.

Battery capacity = 813 Ah (for 5kw) i.e. rated with 12V 813Ah

Battery capacity = 980 Ah (for 6kw) i.e. rated with 12V 980 Ah

5.6.1.3 Converter

The main purpose of the convertor is used to convert the DC current from PV system to AC Current and it has rated efficiency of 85% then,

$$\text{Converter} = \text{Parray} \times 100/85$$

$$\text{Converter} = 6\text{Kw} \times 100/85$$

$$= 7.056 \text{ say } 7 \text{ KW.}$$

The convertor is used to supply the determined load and share the peak load between the battery to satisfy the excess load.

5.6.2 Biogas Energy Potential in Farm

Biogas is mainly produced by the anaerobic digestion of decomposable material or organic waste material which include animal waste, human waste, agricultural crops and other biomass material. The calorific value of the biogas is 20MJ/m^3 (4700Kcal/m^3) and methane is responsible for the energy obtained from the bio gas. There are different models of Biogas system available: KVIC model, Janta model and Dhenabandhu model. Mostly KVIC model is preferred for community or institutional type which is a floating drum type Biogas plant. To identify the correct the sizing of the Biogas plant the following steps should be followed.

5.6.2.1 *Volume Calculation of Digester*

No of cows in the farm = 10 (body weight of 200kg each)
Considering the HRT = 40 days
Total discharge (cow dung) per day = 10kg × 10 cows
= 100kg cow dung is available every day.
Total Solids TS of fresh discharge = 100kg×0.16 = 16 kg.

In 8% concentration of TS

8Kg of solid = 100 kg of influent
1kg of solid = 100/8 Kg of influent
16 kg of TS = 100×16/8 = 200 kg

Volume of water required for the slurry to make the discharge of 8% concentration of TS = 200kg-100kg = 100kg, which means that 100lt water is mixed with the 100 kg of cow dung.

Total slurry per day is 200 l/day
Volume of the digester = V gs (Volume of gas storage chamber) × V_f (volume of the fermentation chamber) = Q×HRT
= 200l/day × 40=8000 l = 8m^3.

5.6.2.2 *Volume of Gas Collecting Chamber (Vc)*

Table 5.4 shows the standard composition of Biogas produced from the Biogas plat at proper blending mixing ratio.

Assuming 1 kg cow dung produces 0.034 m^3 biogas after its retention time then, 200 kg of cow dung will produce 100×0.034 = 3.4 m^3

Table 5.4 Composition of biogas.

Substance	Symbol	Percentage (%)
Methane	CH_4	50-70
Carbon Dioxide	CO_2	30-70
Hydrogen	H_2	5-10
Nitrogen	N_2	1-2
Water Vapour	H_20	0-3
Hydrogen Sulphide	H_2S	Traces

Total size of the Biogas plant = Volume of digester + Volume of gas collecting chamber
= 8+3.4 = 11.4 (say $12m^3$)
The Required plant size is $12m^3$

5.6.2.3 *Generator Sizing*

Considering $0.8m^3$ of biogas is utilized for 1kw then $3.4m^3/0.8m^3$ = 4.25kWh /day (Generator running for 4 hours)
Capacity of the Generator = 4.2kWh/ 4 = 1.06 kW (say 2 KW)
Pf of the generator is 0.8 then 2/0.8 = 2.5 KVA

While sizing the biogas plant for the required quantity of waste, Volume of digester, volume of gas collecting chamber and Generator is very important.

5.6.3 Biomass Potential in the Particular Site

Biomass gasification is a widely used technology ranging from KW (small-scale) to MW (large-scale) application to convert waste material into energy. Biomass is available raw material in the rural areas for the production of electricity. Gasifier is divided into fixed bed and fluidized bed. In India the usage of fixed bed downdraft gasifier is common and traffic rate of electricity generation varies from state to state 3.5KWh to 5 KWh.

Assumption for Initial Design conditions

Type of Biomass – wood type of Gasifier - Downdraft (with open at the top)

Calorific Value - 19.85MJ/Kg
Density of the feed stock - 225kg/m3
Gasification efficiency - 65%
Thermal output (P) - 8KW
Low Heating value of the fuel - 4000KJ
Angle of Inclination of air Nozzle is 10 to 25. For this study the 20 inclination is used.

5.6.3.1 *Syn Gas Generation Rate*

To determine SG for power output of 8kw

SG = P/LHV
p = power output
= 7.2 m^3/hr.

5.6.3.2 *Fuel Consumption Rate (FCR)*

FCR = P/CV× η
= 2.411kg/hr.
Gasifier operated for 5 hours per day = 2.411×5 = 12.0 kg/day.

For continuous supply of syn-gas for small-scale application Downdraft gasifier is preferred, wood chunks are used at the rate of 1.25kg/hr for stable flame for 50 minutes and wood shaving should be admitted at the rate of 15 minutes per kg for stable flame. This signals the internal combustion engines to produce electricity at a required output [2].

5.7 System Design and Analysis

This paper analyses the two cases which use HOMER (Hybrid Optimization of Model for Electric Renewable) software developed by the NREL (National Renewable Energy Laboratory). Several modeling software programs are available in the market on the analysis of hybrid renewable energy system [3–7] but the HOMER software is the most suitable for modeling and optimizing the results. It is mainly used for feasibility studies and evaluates the available resources at the particular site. Cost analysis of a renewable energy system is also carried out in the HOMER software. Figure 5.5 shows that Layout of the HOMER software from the input side to the output side. Table 5.5 shows that component unit cost of the PV module, Battery, Convertor,

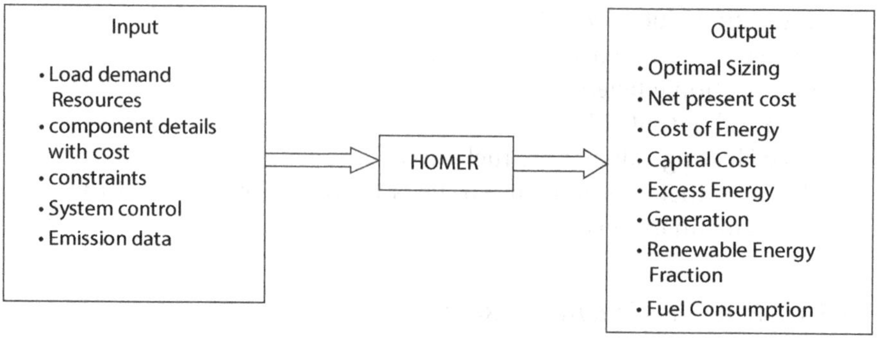

Figure 5.5 Layout of HOMER Software [9].

Table 5.5 Component unit cost (Considering 1$ = Rs.66).

Component	Unit cost ($/ KW)	Replacement ($/KW)	Operation and maintenance cost ($/h) per KW	Life of unit	Reference
PV Module	952	760	10	25	–
Battery	390	365	20	10	–
Convertor	225	177	2.5	25	–
Biomass generator	694	625	0.016	10	[3]
Biogas Generator	1940	1746	0.362	10	[3]

Biogas Generator, Biomass Generation (depending upon the requirement of the case studies) along with their operation and maintenance cost.

5.7.1 Result Analysis

The techno-economical aspects of the present system with suitable variable or constraints which affect the operating cost and output of the system were studied. In sensitivity and optimization analysis the biomass supply and solar radiation is varied to get the COE (Cost of Energy), Net Present Cost and Operating cost of the system. Price of the Biomass $=0.30t/day is considered in the system. By varying the size of the battery, sensitivity of the system is analyzed. The hybrid system is simulated by the HOMER software by varying the PV system and biogas generator capacity to identify the best optimized and economical system for the particular farm.

Small-scale load requirement of the farm is satisfied by this Hybrid system studied in the HOMER software.

5.7.1.1 Case-1 PV/Biomass Hybrid System

Figure 5.6 shows that PV/Biomass Hybrid System shows the schematic model of the hybrid system. Figure 5.12 shows that PV/Biogas Hybrid System shows the schematic model of the hybrid system. It has been found that Total Electricity production from the Hybrid system is 8182KWhr/yr, out of which 7375KWh/yr from PV and 808 KWh/yr. Total AC primary load consumption is 3675KWh/yr and Excess electricity is 53%. From Optimal analysis the best suitable Hybrid system 6KW of Biomass and 4KW of PV with COE (Cost of Energy) is INR 17.51, as shown in Figure 5.7. The capital cost, Replacement cost and Operation and Maintenance cost from the software analysis found that the overall system cost INR 9 Lakhs for a period of 20 years as shown in Figure 5.8. Due to the resources available at the particular site, the cost of PV and battery is a major portion. Biogas resource is available throughout the year as shown in Figure 5.9 at a particular location. The graph is plotted between the Power output of PV and Biogas Generator vs. Days as shown in Figure 5.10, and Figure 5.11 shows the utilization of Resources in an efficient manner.

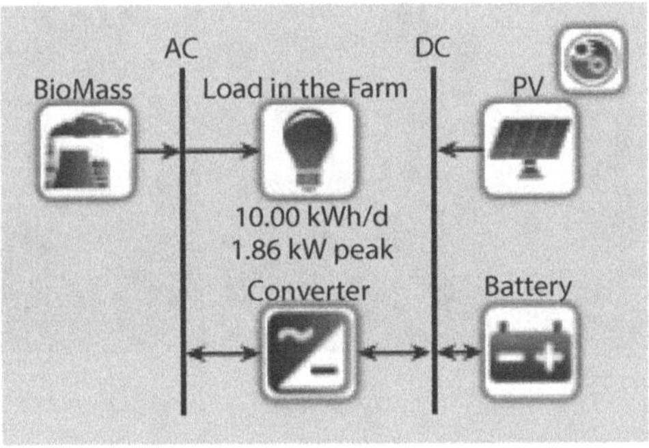

Figure 5.6 Schematic model of the Case-1 PV/Biomass Hybrid system.

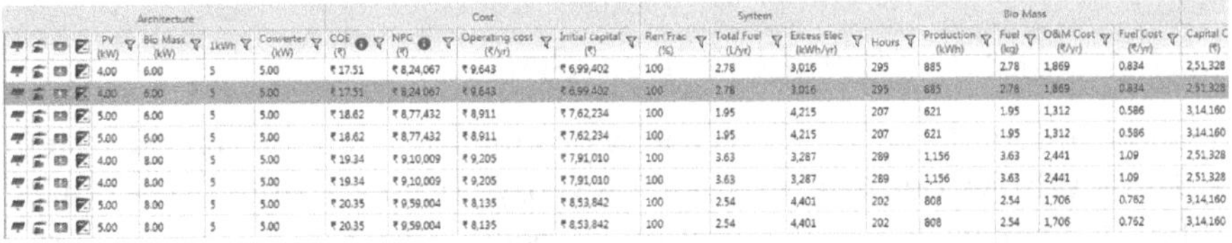

Architecture				Cost				System			Bio Mass					
PV (kW)	Bio Mass (kW)	1kWh	Converter (kW)	COE (₹)	NPC (₹)	Operating cost (₹/yr)	Initial capital (₹)	Ren Frac (%)	Total Fuel (L/yr)	Excess Elec (kWh/yr)	Hours	Production (kWh)	Fuel (kg)	O&M Cost (₹/yr)	Fuel Cost (₹/yr)	Capital C (₹)
4.00	6.00	5	5.00	₹17.51	₹8,24,067	₹9,643	₹6,99,402	100	2.78	3,016	295	885	2.78	1,869	0.834	2,51,328
4.00	6.00	5	5.00	₹17.51	₹8,24,067	₹9,643	₹6,99,402	100	2.78	3,016	295	885	2.78	1,869	0.834	2,51,328
5.00	6.00	5	5.00	₹18.62	₹8,77,432	₹8,911	₹7,62,234	100	1.95	4,215	207	621	1.95	1,312	0.586	3,14,160
5.00	6.00	5	5.00	₹18.62	₹8,77,432	₹8,911	₹7,62,234	100	1.95	4,215	207	621	1.95	1,312	0.586	3,14,160
4.00	8.00	5	5.00	₹19.34	₹9,10,009	₹9,205	₹7,91,010	100	3.63	3,287	289	1,156	3.63	2,441	1.09	2,51,328
4.00	8.00	5	5.00	₹19.34	₹9,10,009	₹9,205	₹7,91,010	100	3.63	3,287	289	1,156	3.63	2,441	1.09	2,51,328
5.00	8.00	5	5.00	₹20.35	₹9,59,004	₹8,135	₹8,53,842	100	2.54	4,401	202	808	2.54	1,706	0.762	3,14,160
5.00	8.00	5	5.00	₹20.35	₹9,59,004	₹8,135	₹8,53,842	100	2.54	4,401	202	808	2.54	1,706	0.762	3,14,160

Figure 5.7 Optimization results from HOMER software.

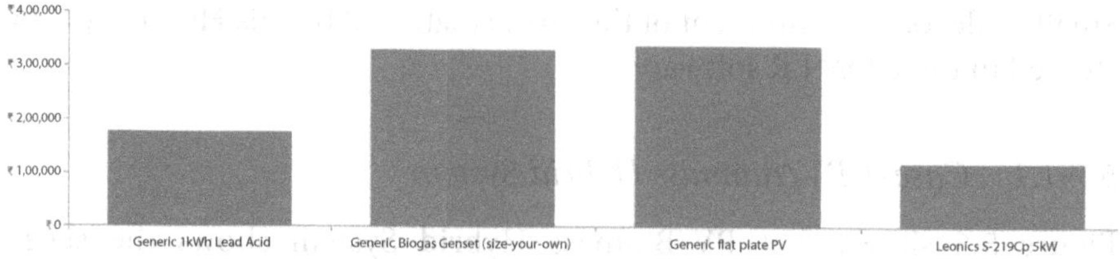

Component	Capital (₹)	Replacement (₹)	O&M (₹)	Fuel (₹)	Salvage (₹)	Total (₹)
Generic 1kWh Lead Acid	₹ 99,000.00	₹ 87,460.33	₹ 646.38	₹ 0.00	₹ -11,858.11	₹ 1,75,248.59
Generic Biogas Genset (size-your-own)	₹ 3,66,432.00	₹ 0.00	₹ 22,060.76	₹ 9.85	₹ -59,092.93	₹ 3,29,409.67
Generic flat plate PV	₹ 3,14,160.00	₹ 0.00	₹ 21,330.40	₹ 0.00	₹ 0.00	₹ 3,35,490.40
Leonics S-219Cp 5kW	₹ 74,250.00	₹ 51,601.59	₹ 0.00	₹ 0.00	₹ -6,996.29	₹ 1,18,855.30
System	₹ 8,53,842.00	₹ 1,39,061.92	₹ 44,037.53	₹ 9.85	₹ -77,947.33	₹ 9,59,003.97

Figure 5.8 Total net present cost of the component for the Hybrid system.

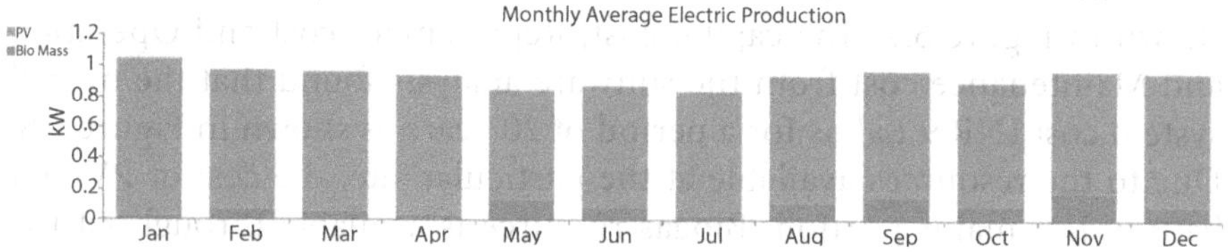

Figure 5.9 Monthly average electricity Production PV/ biomass hybrid system.

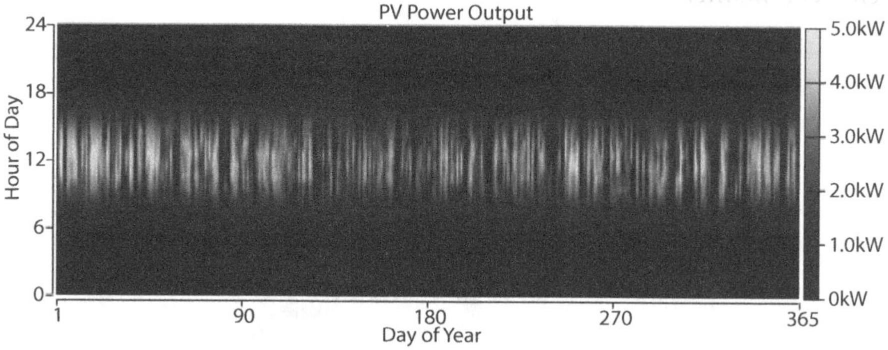

Figure 5.10 PV power output vs days.

5.7.1.2 *Case 2 – Hybrid PV/Biogas System*

It has been found that Total Electricity production from the Hybrid system is 9136KWhr/yr out of which 8849 KWh/yr from PV and 287 KWh/yr. Total AC primary load consumption is 3664 KWh/yr and Excess electricity is 58 %. From Optimal analysis the best suitable Hybrid system 6KW

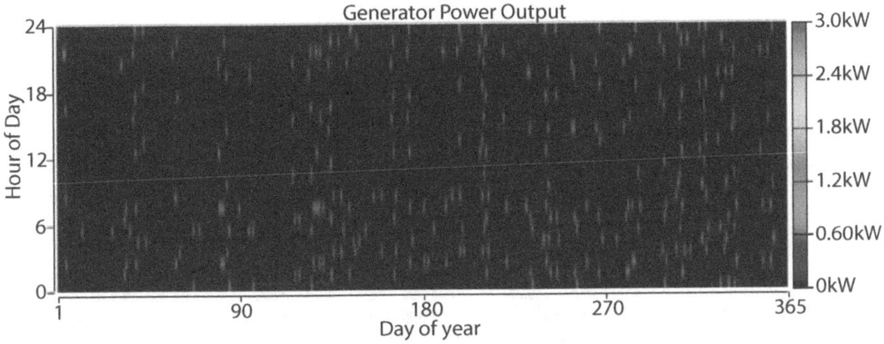

Figure 5.11 Biomass generator power output vs days.

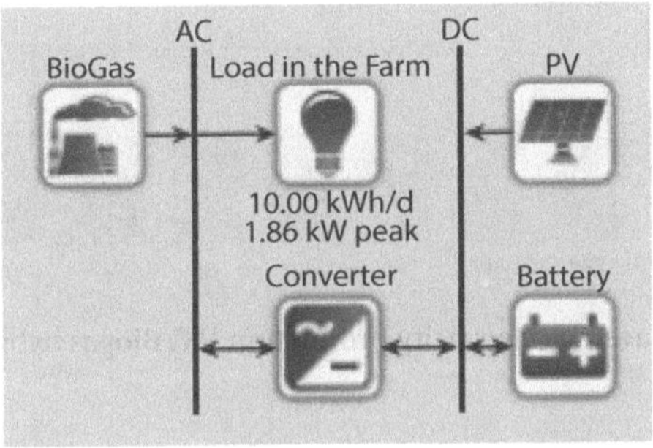

Figure 5.12 Schematic model of the Case-2 PV/Biogas Hybrid system.

of Biomass and 4KW of PV with COE (Cost of Energy) is INR 27.27 as shown in Figure 5.13. The capital cost, Replacement cost and Operation and Maintenance cost of the individual components from the software analysis found overall system cost of INR 12 Lakhs for a period of 20 years as shown in Figure 5.14. Due to the resources available at the particular site, cost of PV and Biogas Generator has been a major portion only for 4 months as shown in Figure 5.15. Usage of PV and Biogas generator is

Architecture				Cost				System			Bio Gas				
PV (kW)	Bio Gas (kW)	Battery	Converter (kW)	COE (₹)	NPC (₹)	Operating cost (₹/yr)	Initial capital (₹)	Ren Frac (%)	Total Fuel (L/yr)	Excess Elec (kWh/yr)	Hours	Production (kWh)	Fuel (kg)	O&M Cost (₹/yr)	Fuel Cost (₹/yr)
6.00	4.00	5	5.00	₹ 27.27	₹ 1.29M	₹ 17,235	₹ 1.06M	100	0.899	5,347	136	287	0.899	12,996	0.270
6.00	4.00	5	5.00	₹ 27.27	₹ 1.29M	₹ 17,235	₹ 1.06M	100	0.902	5,348	136	288	0.902	12,996	0.271
4.00	4.00	5	5.00	₹ 28.54	₹ 1.34M	₹ 31,449	₹ 9,36,738	100	1.82	2,706	276	582	1.82	26,375	0.546
4.00	4.00	5	5.00	₹ 28.54	₹ 1.34M	₹ 31,449	₹ 9,36,738	100	1.83	2,708	276	584	1.83	26,375	0.548
2.00	4.00	5	5.00	₹ 48.33	₹ 2.28M	₹ 1,13,271	₹ 8,11,074	100	6.72	1,370	1,023	2,148	6.72	97,758	2.02
2.00	4.00	5	5.00	₹ 48.33	₹ 2.28M	₹ 1,13,271	₹ 8,11,074	100	6.75	1,379	1,023	2,158	6.75	97,758	2.03

Figure 5.13 Optimization results from HOMER software for PV/Biogas.

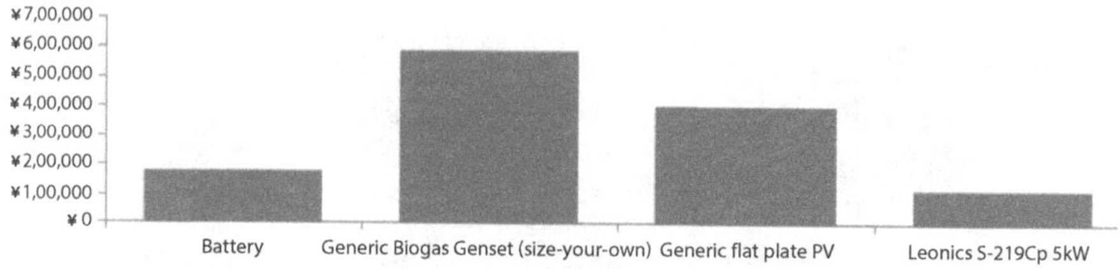

Component	Capital (₹)	Replacement (₹)	O&M (₹)	Fuel (₹)	Salvage (₹)	Total (₹)
Battery	₹ 99,000.00	₹ 87,460.33	₹ 646.38	₹ 0.00	₹ -11,858.11	₹ 1,75,248.59
Generic Biogas Genset (size-your-own)	₹ 5,12,160.00	₹ 0.00	₹ 1,68,008.07	₹ 3.49	₹ -91,650.89	₹ 5,88,520.67
Generic flat plate PV	₹ 3,76,992.00	₹ 0.00	₹ 25,596.48	₹ 0.00	₹ 0.00	₹ 4,02,588.48
Leonics S-219Cp 5kW	₹ 74,250.00	₹ 51,601.59	₹ 0.00	₹ 0.00	₹ -6,996.29	₹ 1,18,855.30
System	₹ 10,62,402.00	₹ 1,39,061.92	₹ 1,94,250.93	₹ 3.49	₹ -1,10,505.29	₹ 12,85,213.05

Figure 5.14 Total net present cost of the component for the Hybrid system.

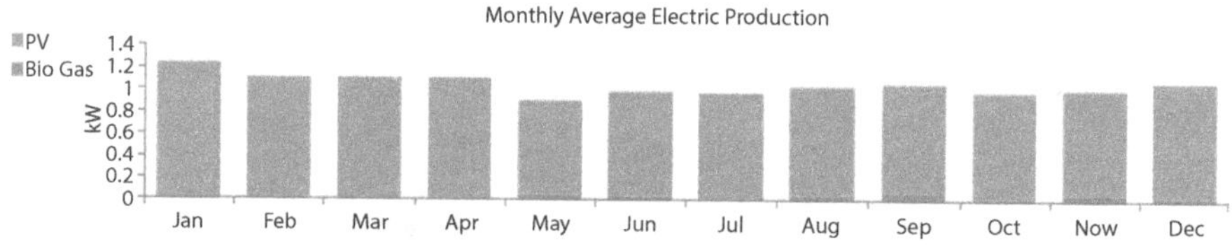

Figure 5.15 Monthly average electricity production PV/Biogas hybrid system.

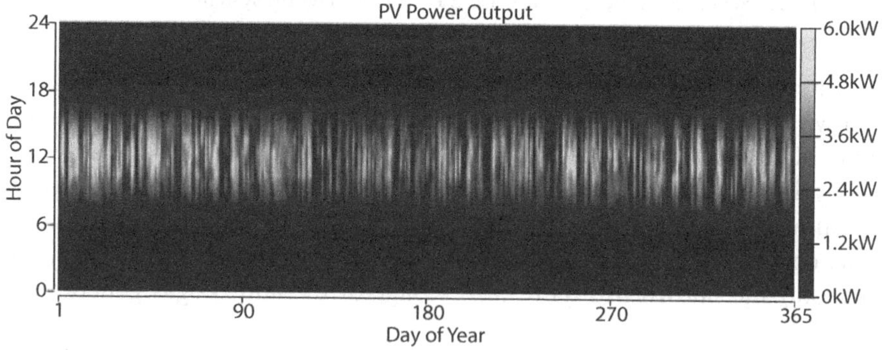

Figure 5.16 PV power output vs. days.

plotted in the graph which shows utilization of the system in KW vs. Days as shown in Figure 5.16 & Figure 5.17. Comparison of the economics of the two simulated model of the PV/Biogas hybrid system with payback period of 3 years is shown in Figure 5.18.

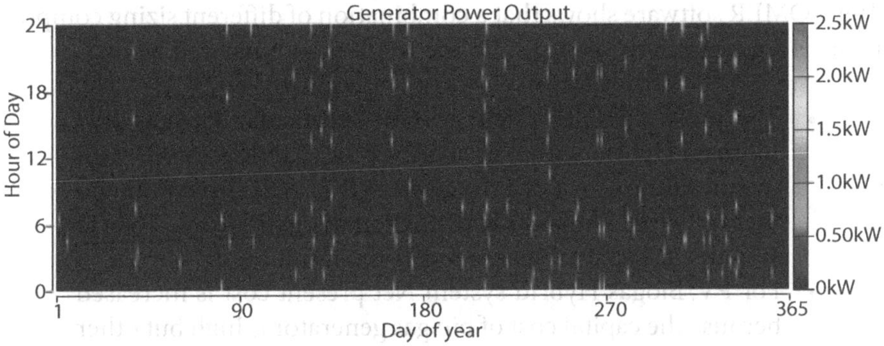

Figure 5.17 Biogas generator power output vs. days.

	Architecture									Cost	
						PV (kW)	Bio Gas (kW)	Battery	Converter (kW)	NPC (₹)	Initial capital (₹)
Base system						2.00	4.00	5	5.00	₹ 2.28M	₹ 8,11,074
Current system						6.00	4.00	5	5.00	₹ 1.29M	₹ 1.06M

Metric	Value
Present worth (₹)	₹ 9,90,176
Annual worth (₹/yr)	₹ 76,594
Return on investment (%)	37.3
Internal rate of return (%)	33.4
Simple payback (yr)	3.01
Discounted payback (yr)	3.42

Figure 5.18 Comparison of current and base system.

5.8 Conclusion

The proposed model shows that a Hybrid system has enough potential to power the farm without external supply of grid connection. This type of system is used for remote villages and farms which are isolated from the grid connection. The excess energy generated from the Renewable energy system is sold to the nearby village and wastage from the biogas plant is dried, which can be sold to the agricultural farm or in the form of vermin compost. This study shows that a Hybrid system is environmentally feasible and implementation in a farm can be encouraged. The government has to prepare a new policy which will give subsidies to these schemes to encourage low-income farmers to implement this type of system.

The HOMER software shows that a combination of different sizing components is analyzed in the system and the result is summarized below:

- The most suitable system for this particular farm is PV/Biomass Hybrid with Net present cost of INR 8,24,067 with the minimum Cost of Energy as INR 17.8. It is recommend that 6 KW of PV and 6KW of Biomass gasification Hybrid system can satisfy the demand of the farm.
- For PV/Biogas Hybrid system Net present cost is increased because the capital cost of bio gas generator is high but other income is obtained by Excess Electricity generated system from the PV sold at a rate INR 5.4 KW/h and Vermi compost can be sold at INR12 per Kg.
- The highest net Present cost of INR 20 lakhs and Cost of Energy is INR 48.33 is obtained for the 2KW of PV and 4KW of Biogas.
- From the Hybrid Renewable Energy system the Carbon emission is low when compared to the Conventional system of electricity generation and this makes a positive impact on the environment and society. It leads to independence from the grid system.

References

1. S. Appassmandri *et al.*, Production and Profitability Analysis of Grapevine Orchard in Coimbatore, Tamil Nadu, India, *International Journal of Current Microbiology and Applied Sciences*, vol. 6, pp. 2072-2180, 2017.
2. P.E. Akhator *et al.*, Design and Development of Small-Scale Biomass Downdraft gasifier, *Nigerian Journal of Technology*, vol. 38, No. 4, pp. 922-930, 2019.
3. Sonali Goel, Renu Sharma, Optimal sizing of biomass-biogas hybrid system for sustainable power supply to a commercial agricultural farm in northern Odisha, India, *Environ Dev Sustain.*, 2018.
4. Valente, L. and de Almeida, Economic analysis of a diesel/photovoltaic hybrid system for decentralized power generation in northern Brazil. *Energy*, 23(4), pp. 317-323, 1998.
5. Lau, K., Yousof *et al.*, Performance analysis of hybrid photovoltaic/diesel energy system under Malaysian conditions. *Energy*, 486 35(8), pp. 3245-3255, 2010.
6. Olatomiwa Lanre, Mekhilef Saad, Huda ASN, Ohunakin S, Economic evaluation ofhybrid energy systems for rural electrification in six geo political zones of Nigeria. *Renew Energy*; 83:435–46, 2015.

7. Bhatt, A., Sharma, M. and Saini, R, Feasibility and sensitivity analysis of an off-grid micro hydro–photovoltaic–biomass and biogas–diesel–battery hybrid energy system for a remote area in Uttarakhand state, India. *Renewable and Sustainable Energy Reviews*, 61, pp. 53-69, 2016.
8. Hossam S *et al.*, Design and Optimization of a Grid-Tied PV-Biomass Hybrid Renewable Energy with battery storage: A case study for a small building in Hurghada (Egypt), *International Journal of Scientific & Technology Research*, Vol. 8, pp. 2282-2289, 2019.
9. Sunanda Sinha, S.S. Chandel, Review of software tools for hybrid renewable energy system. *Renewable and Sustainable Energy Review*, vol. 32, pp. 192-205, 2014.
10. S. Bhramara, M. Paras Moghaddam, M.R. Haghifam, Optimal Planning of hybrid renewable energy systems using HOMER: A review, *Renewable and Sustainable Energy Review* 62, 2016.
11. J.G. Castellanos *et al.*, Modeling an off-grid integrated renewable energy system for rural electrification in India using photovoltaic and anaerobic digestion, *Renewable Energy*, 74, pp. 390-398, 2015.
12. NREL, available at https://nsrdb.nrel.gov/
13. NIWE, available at https://niwe.res.in/department_srra.php

6

Experimental Investigation of Solar Photovoltaic Cold Storage With Thermal Energy Storage

K. Sahoo[1], V. Yadav[2*], N. Goyal[3], S. Kumar[4], Y. Singh[2], S. Mukhopadhyay[5], U. Sahoo[2], A.K. Tripathi[2] and C. Banerjee[2]

[1]Indian Institute of Technology Kharagpur, Kharagpur, India
[2]National Institute of Solar Energy, Gwal Pahari, Gurugram, Haryana, India
[3]Inficold India Private Limited, Ghaziabad, Uttar Pradesh, India
[4]Indian Institute of Technology Delhi, New Delhi, Delhi, India
[5]Indian Institute of Engineering Science and Technology, Shibpur, West Bengal, India

Abstract
The chapter is mainly focused on experimental investigation of solar photovoltaic (SPV) cold storage integrated with thermal energy storage, which can be used for the storage of 10 MT of vegetables in the temperature range of 4°C–10°C. The energy requirement observed for 1 day was around 72 kWh, which is huge for such a small cold storage system. This research presents a SPV-driven cold storage with thermal energy storage as a backup system for non-sunshine hours with autonomy of 20 hours in loading condition and 35 hours in no-load condition. The compressor is driven by a SPV panel having capacity of 3 TR, which is designed with maximum power point tracking (MPPT) controller and provides easy starting, maximum power tracking, and better power management for the system. This results in the proposed compressor operating from early morning to late evening and also operating during the day with low solar irradiance and hence improved the utilization of SPV power. The experimental investigation and economic analysis, i.e., feasibility, IRR, NPV, and payback for different possible business models of the system, has been carried in this chapter.

***Keywords*:** Solar energy, photovoltaic, solar thermal, phase change material, maximum power point tracking, economic feasibility

**Corresponding author*: vikrant.nise@gmail.com

Umakanta Sahoo (ed.) Hybrid Renewable Energy Systems, (135–168) © 2021 Scrivener Publishing LLC

6.1 Introduction

The cold chain system is very crucial for the growth of global trade. Every year, billions of tons of vegetables and food products are wasted due to lack of adequate and proper cold storage facility. Globally, billions of dollars are spent on improvement of agricultural process to create higher food yield, but nearly half of all food never makes it to a consumer. According to United Nations report, 1.3 billion tons of food is wasted annually, causing economic losses of $750 billion [1]. This represents the missed opportunity to improve global food security. The total capacity of refrigerated warehouses worldwide was 616 million cubic meters in 2018, 2.67% greater than the capacity reported in 2016. India was the single largest country market, at 150 million cubic meters, followed by the United States at 131 million cubic meters, and China at 105 million cubic meters. The global cold storage market size was valued at USD 94.02 billion in 2018 and is projected to expand at a CAGR of 12.2% during the forecast period of 2019–2025 [2]. The global cold chain market is expanding with considerable growth potential for over the next 5 years. The growth of this market can be attributed to the growth of international trade of perishable foods, technological advancements in refrigerated storage and transport, government support for the infrastructural development of the cold chain industry, and increase in consumer demand for perishable foods. Refrigerated storage has become an integral part of the Supply Chain Management (SCM) when it comes to transporting and storing temperature-sensitive products.

The estimated annual production of fruits and vegetables in India is about 130 million tons. This accounts for 18% of our agricultural output. Due to diverse agro climatic conditions and better availability of package of practices, the production is gradually rising. Although, there is a vast scope for increasing the production, the lack of cold storage and cold chain facilities are becoming major bottlenecks in tapping the potential. The cold storage facilities now available are mostly for a single commodity like potato, orange, apple, grapes, pomegranates, and flowers, which results in poor capacity utilization. A report by the Institution of Mechanical Engineers (IME), a professional organization of the United Kingdom, states that, in 2018, India has around 6,300 cold storage facilities, with a capacity of 30.11 million tons. However, some 75%–80% of these refrigerated warehouses are suitable only to store potatoes, a commodity that produces only 20% of agricultural revenue. Based on the report, the organization had even emphasized the need for expansion of cold storage infrastructure in an affordable, reliable, and sustainable way to increase the contribution of agriculture to the economy. To develop a solar cold storage with thermal battery, certain parameters need to be considered such as specific outcome, thermodynamic relations, and economic analysis for the system. Thermal energy storage, i.e., thermal battery, is a judicious selection to aid the solar photovoltaic (SPV) cold storage for operation during short transients and night.

In the present study, an attempt is made to achieve efficiency as compared to stand-alone system. The main objective of this study is to experimental investigation for SPV cold storage with thermal storage. The system was evaluated, and observations were recorded on the basis of field performance data to identify the effects of various operating parameters. The payback period of the developed system is also provided in this chapter.

6.2 Scope of Cold Storage in India

Preservation of food has been a serious matter of concern both for the farmers and food industries. India is one of the largest producers of the fruits and vegetables across the world, but per capita availability is not that good because of the loss of the fruits and vegetables during the post harvesting. Furthermore, the significant quantity of the produce deteriorates in the supply chain, before it reaches consumers because of issues in handling, storage, transportation, and the perishable nature of fruits and vegetables [3]. The huge loss of the produce from harvesting to the local market increases the cost of the fruits and vegetables, but the real producers (farmers) are still deprived of the monitorial benefits. The huge loss to the farmers may be avoided by the incorporation of certain measures which can save their produce. One of such techniques is the use of cold storage to improve the shelf life of the products. A cold storage facility accessible to them will go a long way in removing the risk of distress sale to ensure better returns. The demand for the processed food industry has led to the need of cold storage chains in India. IME pointed out the increased need of storage capacity by 40%, as only 10%–11% of fruits and vegetables produced in India use cold storage. The organization also pointed out that India's investment in cold chain is projected to be $15 billion over the next 5 years. While approximately 60% of the total numbers of facilities are located only in Uttar Pradesh, Gujarat, West Bengal, and Punjab, only 4 million of the 104 million tons of fresh produce is transported through a cold chain.

Undoubtedly, cold storage forms the most vital post-harvest horticultural value chain for the downstream food processing industry. Top potato producers of Uttar Pradesh and West Bengal make up 55%–60% of the overall domestic cold storage capacity. The current cold storage capacity in India is about 37–39 MT and there are about 7,645 cold storages in the country with 68% of the capacity being used for potato, while 30% is multi-commodity cold storage [4].

Apart from potato, the multi-commodity cold storages cater to meat and poultry, seafood, dairy products, fruits and vegetables, and pharmaceuticals. In most of the industries, the preservation of agro products is carried by cold storage runs on electricity and other conventional sources which results in higher operational cost and greenhouse gas emission. Solar cold storage with thermal battery is one of the most attractive solutions to overcome these challenges. The geographical location of India is such that most of the location receives good sunshine throughout the year.

As per the data, 5,000 trillion kWh energy is incident over India itself in a year, and most of the location receives solar radiation of about 4–5.5 kWh/m^2/day. In India, on average, 280–300 sunny days are available, and some locations receive good sunshine for whole years, which is a best suitable place for runs of solar energy systems [5–10]. Many of the researchers explained about the hybrid renewable energy systems for various end use applications like power generation, vapor absorption refrigeration cooling system, and desalination for drinking purposes [11–16]. Over the years, researchers have developed various types of cold storage and their performance is investigated for the agricultural products in different climatic conditions. Many researchers have analyzed different types of cold storage for agricultural products in different climatic conditions; but to the best of our knowledge, no work has been carried out on SPV cold storage integrated with thermal battery system. The solar energy is transient in nature, so solar cold storage can only be used during bright sunny days but integration with the thermal energy storage can improve the usability of the cold storage during off-sunshine hours and hazy climatic condition. The concept is most important for continuous operation of cold storage for fulfilling the preservation demand.

6.3 Materials and Method

6.3.1 Experimental Setup

The current emphasis of the project lies on the fabrication, testing, and performance analysis of a VCRS-based cold storage for 10 MT of potatoes, which will be powered by SPV panels. The VCRS-based cold storage has been tested in both the seasons and gives almost the same COP of around 3.5. In order to avoid the usage of battery storage for providing the energy backup in the nighttime, an ice thermal energy storage prototype has been designed to be integrated with the current setup. The ice thermal energy storage is designed to provide the energy backup of around 16 hours (in the worst-case scenario) by storing latent heat in the form of ice by utilizing the excess solar energy from SPV panels. This kind

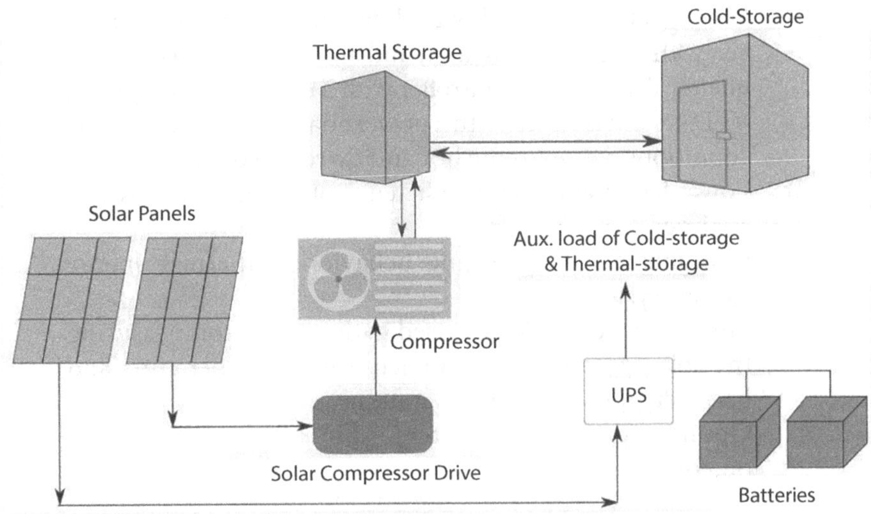

Figure 6.1 Schematic of solar-powered cold storage with thermal storage.

of unique solar-powered cold storage has never been implemented in the region of India. Figure 6.1 represents the schematic of the SPV-powered cold storage.

- **Operating Methodology of the Solar-Powered Cold Storage**

    i. During the daytime, cold room will be run by the SPV. The excess energy stored in the panels during the day will be stored in the form of latent heat energy of thermal energy storage to be utilized in the night.
    ii. In the daytime, two operations will take place simultaneously—one for the charging process of the TES and second for the direct cooling in the cold storage by running the compressor.
    iii. In the nighttime, the energy stored in the TES in the form of solidified ice will be utilized by circulating the Refrigerant R-134a and delivering the cold air through the evaporator of the cold storage.

- **Testing Methodology**

    a) Performance analysis of the cold room is carried out by measuring the various parameters of the cold room under various conditions. The trends of the different parameters were studied through their respective plots.

b) Whole system is tested with different operating condition, i.e., with inverter installation and another with variable frequency drive (VFD) controller with MPPT tracking system.
c) VFD system is tested with a power analyzer for two climatic conditions, i.e., extreme hot and extreme cold conditions.
d) Power factor and energy efficiency of whole system is also analyzed.
e) Charging and discharging behavior of thermal energy storage system.
f) Three thermocouples are used in the system for measuring the cold room temperature, ice storage temperature, and backside of PV panel temperature.
g) Optimization of whole system is presented on the basis of outcome result.

- **Experimental Uncertainty**

The uncertainty (i.e., external uncertainty and internal uncertainty) of most sensitive parameters affecting the SPV cold storage was considered. The external uncertainty is the least count of measuring instruments. The internal uncertainty of SPV panel, crop, and mass was analyzed using the following equations given by [17, 18].

Percentage of internal uncertainty $U_i(\%)$ can be written as follows:

$$U_i(\%) = \frac{U_i}{\textit{mean of total observations}} \times 100$$

where U_i is the internal uncertainty, expressed as follows:

$$U_i = \frac{\sqrt{\sigma_1^2 + \sigma_2^2 + \ldots \sigma_n^2}}{N}$$

where σ is standard deviation, written as follows:

$$\sigma = \frac{\sqrt{\sum (X - \bar{X})^2}}{\mathrm{N_0}}$$

where $(X - \bar{X})^2$ is deviation of observation from mean and N and N_0 are the number of sets and number of observations in each state, respectively.

Percentage of total uncertainty $U(\%)$ is expressed as follows:

$$U(\%) = U_i(\%) + U_e(\%)$$

where $U_e(\%)$ is the external uncertainty.

The experimental analysis of the SPV-based cold storage using thermal energy storage has been carried out by installing the system which consists of 10-MT capacity of vegetable preservation, 3 kW–based VFD compressor, 1 ton capacity ice thermal storage at NISE, Gurugram, Haryana, India (latitude: 28° 25′31.2″N; longitude: 77° 9′18.8″E) shown in Figures 6.2 to 6.5. This system has been evaluated, and observations were recorded on the basis of field data collected as shown in Table 6.1. The energy required to compressor at average ambient temperature on monthly basis is shown in Table 6.2.

6.4 Economic Analysis

Renewable energy faces a challenge which is more economic than technical, the higher initial investment costs as compared to conventional energy sources. This demands an in-depth analysis of the economics of the project. This helps in developing innovative financing methods and also removes uncertainties

Figure 6.2 Thermal energy storage system for providing backup for cooling in cold storage at NISE.

Figure 6.3 Variable frequency drive controller at NISE.

Figure 6.4 Testing of variable frequency drive controller through SPV power analyzer at NISE.

Figure 6.5 Solar powered cold storage installed at NISE.

from the minds of investors [19]. Even though the thermodynamic analysis of the system can be used to improve the efficiency, the feasibility of applying these improvements is generally constrained by the limitations of the financial resources. The cost analysis is to find the actual size that gives the cheapest combination of SPV cold storage integrated with thermal energy storage system. The net present value (NPV) is used in capital budgeting to analyze the profitability of an investment or project. There are two steps involved in the calculation of the NPV. In the first step, the value of net cash inflows from the chapter is calculated. They are calculated by knowing the expected earnings from the project for a certain time period and subtracting the expected or calculated expenditures during the same time period. The net cash inflows may be fixed or variable, and the calculation of the NPV changes accordingly.

When the cash inflows are variable, the NPV of cash flows is written as follows:

$$\mathrm{NPV} = \sum_{T-1}^{T} \frac{\text{Total Cash Flow}}{(1+\mathrm{i})^{\mathrm{T}}} - \text{Initial Investment}$$

where
"i" is expected rate of return.
"T" is the total project life.

Table 6.1 Experimental data of cold storage using VFD-based compressor.

Time	Amb. temp. (°C)	GHI (W/m²)	PV output (kW)	Comp. load (kW)	Auxiliary load (kW)	Grid export (kW)	Cooling output (kWh)	CR load (kWh)	Ice storage (kWh)
5th November 2018									
08:00	21.1	278	2.464	1.9	0.3	Nil	6.65	2.60	4.05
09:00	23.7	452	3.931	3	0.3	0.631	12	2.68	9.32
10:00	27.2	587	4.943	3	0.3	1.643	12	2.76	9.24
11:00	29.6	661	5.515	3	0.3	2.215	12	2.82	9.18
12:00	30.9	667	5.513	3	0.3	2.213	12	2.84	9.16
13:00	31	602	5.032	3	0.3	1.732	12	2.84	9.16
14:00	30.4	475	4.013	3	0.3	0.713	12	2.83	9.17
15:00	29.3	305	2.576	2.03	0.3	Nil	7.1	2.82	4.28
4th December 2018									
08:00	14.5	242	2.161	1.7	0.3	Nil	5.95	2.42	3.53
09:00	18.3	383	3.731	3	0.3	0.431	12	2.55	9.45
10:00	22.6	517	4.901	3	0.3	1.601	12	2.65	9.35

(Continued)

Table 6.1 Experimental data of cold storage using VFD-based compressor. (*Continued*)

Time	Amb. temp. (°C)	GHI (W/m²)	PV output (kW)	Comp. load (kW)	Auxiliary load (kW)	Grid export (kW)	Cooling output (kWh)	CR load (kWh)	Ice storage (kWh)
11:00	24.5	595	5.425	3	0.3	2.125	12	2.69	9.31
12:00	25.3	608	5.499	3	0.3	2.199	12	2.71	9.29
13:00	25.3	552	5.051	3	0.3	1.751	12	2.71	9.29
14:00	24.5	437	4.109	3	0.3	0.809	12	2.69	9.31
15:00	22.9	299	2.642	2.08	0.3	Nil	7.28	2.66	4.62
24th January 2019									
08:00	12.7	230	1.829	1.4	0.3	Nil	4.9	2.39	2.51
09:00	17.6	373	3.661	3	0.3	0.361	12	2.53	9.47
10:00	21.6	531	4.998	3	0.3	1.698	12	2.63	9.37
11:00	23.3	628	5.717	3	0.3	2.417	12	2.65	9.35
12:00	24.2	671	6.049	3	0.3	2.749	12	2.69	9.31
13:00	24.3	641	5.793	3	0.3	2.493	12	2.69	9.31
14:00	23.6	541	5.005	3	0.3	1.705	12	2.68	9.32

(*Continued*)

Table 6.1 Experimental data of cold storage using VFD-based compressor. (*Continued*)

Time	Amb. temp. (°C)	GHI (W/m²)	PV output (kW)	Comp. load (kW)	Auxiliary load (kW)	Grid export (kW)	Cooling output (kWh)	CR load (kWh)	Ice storage (kWh)
15:00	22.3	392	3.75	3	0.3	0.45	12	2.65	9.35
14th February 2019									
08:00	17.5	252	2.192	1.7	0.3	Nil	5.95	2.53	3.42
09:00	22.5	438	3.89	3	0.3	0.59	12	2.65	9.35
10:00	25.3	605	5.128	3	0.3	1.828	12	2.71	9.29
11:00	26.8	717	5.901	3	0.3	2.601	12	2.75	9.25
12:00	27.5	758	6.143	3	0.3	2.843	12	2.77	9.23
13:00	27.6	730	5.945	3	0.3	2.645	12	2.77	9.23
14:00	26.9	629	5.23	3	0.3	1.93	12	2.76	9.24
15:00	25.4	471	4.055	3	0.3	0.755	12	2.71	9.29
10th March 2019									
09:00	30.1	531	4.224	3	0.3	0.924	12	2.83	9.17
10:00	32.1	700	5.418	3	0.3	2.118	12	2.88	9.12

(*Continued*)

Table 6.1 Experimental data of cold storage using VFD-based compressor. (*Continued*)

Time	Amb. temp. (°C)	GHI (W/m²)	PV output (kW)	Comp. load (kW)	Auxiliary load (kW)	Grid export (kW)	Cooling output (kWh)	CR load (kWh)	Ice storage (kWh)
11:00	29.2	310	3.312	3	0.3	0.012	7.9	2.82	5.08
12:00	33.6	859	6.456	3	0.3	3.156	12	2.92	9.08
13:00	33.4	823	6.306	3	0.3	3.006	12	2.91	9.09
14:00	32.8	713	5.556	3	0.3	2.256	12	2.9	9.1
15:00	31.5	547	4.292	3	0.3	0.992	12	2.87	9.13
13th April 2019									
09:00	34.7	661	4.737	3	0.3	1.437	12	2.94	9.06
10:00	37.1	821	5.744	3	0.3	2.444	12	2.98	9.02
11:00	38.8	915	6.292	3	0.3	2.992	12	2.99	9.01
12:00	39.6	938	6.301	3	0.3	3.001	12	3	9
13:00	39.6	890	5.998	3	0.3	2.698	12	3	9
14:00	24.8	350	3.59	3	0.3	0.29	8.8	2.52	6.28
15:00	25.2	385	3.7	3	0.3	0.4	8.9	2.7	6.2

Table 6.2 Energy required to compressor at average ambient temperature on monthly basis.

Month	Average ambient temperature (°C)	Energy required for compressor (kWh)
January	13	19.36
February	17	20.16
March	22	21.12
April	29	22.48
May	33	23.2
Jun	32	23.04
July	31	22.8
August	30	22.64
September	29	22.48
October	25	21.68
November	20	20.76
December	15	19.76

Conditions:

1. If NPV is zero or positive (for the project lifetime), the project can be accepted. However, it is advisable to have the NPV positive rather than zero, as NPV is calculated on future estimated cash flows. If there are changes in the cash flows due to malfunction of system, there is a risk of a zero NPV changing to a negative NPV. It is the discount rate often used in capital budgeting that makes the NPV of all cash flows from a particular project equal to zero. Thus, it can be represented as that value of "i" for which NPV = 0. Thus, we ge

$$\mathrm{NPV} = \sum_{T-1}^{T} \frac{\text{Total Cash Flow}}{(1+\mathrm{i})^{\mathrm{T}}} - \text{Initial Investment} = 0$$

Calculating the IRR involves assumption of a rate of return "i" and calculating the corresponding NPV. This process has to be repeated till the NPV is zero, and the rate of return used for this calculation is the Internal Rate of Return. The higher the Internal Rate of Return, the more lucrative is the project. However, it is imperative that the IRR is higher than the cost of securing the funds. Also, it should be sufficiently higher than other comparable investment options.

2. If NPV is negative, the project is not fulfilling the expectations for profitability. However, in case of social or public sector projects, negative NPV is often observed. This is because in such projects, making a profit is not the aim of the project, and a lower rate of return may be incurred over the lifetime of the project.

6.4.1 Payback Period

Payback period is the time in which the initial cash outflow of an investment is expected to be recovered from the cash inflows generated by the investment. It is simply the ratio of initial investment to annual cash flow.

- **Assumptions while doing economic analysis**

  a. Operation and maintenance cost: 1.5% of project cost escalating at 10% per year
  b. Project life is 15 year
  c. Cost of electricity and heat: Calculated for electricity and diesel as fuel.
  d. Electricity tariff: Average of commercial tariffs in Delhi (average Rs. 6/unit with an annual increment of 2.5%).
  e. Bank interest is taken as 6% per annum.

The total investment costs of SPV and solar thermal–powered cold storage are given in Tables 6.3 and 6.4.

6.5 Different Business Models for SPV Cold Storage With Thermal Energy Storage

Model-1: If a Conventional Cold Storage is Replaced by SPV Cold Storage.

The installation cost of SPV cold storage and NPV and IRR analysis of Model-1 are given in Tables 6.5 and 6.6.

Table 6.3 Installation cost of solar photovoltaic cold storage.

SL. No.	Components	Cost (in Lakhs)
1.	Cold Room	2.5
2.	Refrigeration Unit	1.2
3.	Thermal Storage Unit	2.5
4.	9.5 kWp Solar Panel	3.8
5.	Solar Controller	0.4
6.	Solar Inverter	0.15
7.	Battery	0.6
8.	Installation	0.25
9.	Tax Burden	0.6
Total Amount		12

Table 6.4 Installation cost of solar thermal cold storage.

SL. No.	Components	Cost (in Lakhs)
1.	Cold Room	2.5
2.	Refrigeration Unit	4.55
3.	Thermal Storage Unit	2.5
4.	Solar Collector (PTC)	4.64
5.	2 kWp Solar Panel (Auxiliary Load)	0.8
6.	Solar Inverter	0.15
7.	Battery	0.6
8.	Installation	0.5
9.	Tax Burden	0.76
Total Amount		17

Table 6.5 Installation cost of SPV cold storage.

SL. No.	Components	Cost (in Lakhs)
1.	Thermal Storage Unit	2.5
2.	9.5 kWp Solar Panel	3.8
3.	Solar Controller	0.4
4.	Solar Inverter	0.15
5.	Battery	0.6
6.	Installation	0.25
7.	Tax Burden	0.6
Total Amount		8.3

Model-2: If We Install a New SPV Cold Storage for Saving Wastage Fruits and Vegetable.

As per business Model-2, the installation cost of SPV cold storage and NPV and IRR analysis are given Tables 6.7 and 6.8. The solar-based cold storage with thermal battery is assumed for this analysis with weekly loading and unloading of 10 MT of vegetables, i.e., approx. 200 MT of perishable food can be stored annually (assuming the cold storage is running 40 week and each product stored for 2 week). Annual wastage in absence of cold storage = 30% of 200 MT = 60 MT of perishable food.

The 70% of these wastages, i.e., 70% of 60 MT = 42 MT of perishable food, can be saved with the presence of cold storage. Considering the cost saving of Rs.10/kg of vegetables = 42 * 10 = 4.2 lakhs can be saved annually with cold storage. Saving due to transportation, bargaining power to farmers = 160 *1 = 1.6 lakhs (cost saving at 0.5 Rs/kg). The net savings are 5.84 lakhs.

Model-3: If We Install a New SPV Cold Storage on the Basis of Profit Mode.

OPTION-1

If the product stored in rent basis, the rent charge is Rs. 1.2/kg/week (40 weeks in a year).

Collected revenue will be = 10,000*1.8*20 = 3.6 lakhs per year. The NPV and IRR analysis of Model-3 (Option-1) is given in Table 6.9.

Table 6.6 NPV and IRR analysis of Model-1.

Year	Net saving	O & M cost	Bank payment	Net expenditure	Net cash flow	Cumulative net cash flow	NPV
1.	346,129	0	105,133.3	105,133.3	240,995.7	240,995.7	−589,004.3
2.	350,333.8	12,450	101,813.4	114,263.4	236,070.4	477,066.1	−352,933.9
3.	354,538.6	13,695	98,493.33	112,188.3	242,350.3	719,416.3	−110,583.7
4.	359,023.7	15,064	95,173.33	110,237.3	248,786.4	968,202.7	138,202.7
5.	363,508.8	16,570	91,853.33	108,423.3	255,085.5	122,3288	393,288.2
6.	368,274.3	18,228	88,533.33	106,761.3	261,512.9	148,4801	654,801.2
7.	373,039.7	20,050	85,213.33	105,263.3	267,776.4	175,2578	922,577.6
8.	377,805.2	22,055	81,893.33	103,948.3	273,856.8	202,6434	119,6434
9.	382,850.9	24,261	78,573.33	102,834.3	280,016.6	230,6451	147,6451
10.	387,896.7	26,687	75,253.34	101,940.3	285,956.3	259,2407	176,2407
11.	393,222.8	29,356	71,933.34	101,289.3	291,933.4	288,4341	205,4341
12.	398,548.8	32,292	68,613.34	100,905.3	297,643.5	318,1984	235,1984
13.	404,155.2	35,521	65,293.34	100,814.3	303,340.9	348,5325	265,5325
14.	409,761.6	39,073	61,973.34	101,046.3	308,715.3	379,4040	296,4040
15.	415,648.4	42,980	58,653.34	101,633.3	314,015	410,8055	327,8055

Table 6.7 Installation cost of SPV cold storage.

SL. No.	Components	Cost (in Lakhs)
1.	Cold Room	2.5
2.	Refrigeration Unit	1.2
3.	Thermal Storage Unit	2.5
4.	9.5 kWp Solar Panel	3.8
5.	Solar Controller	0.4
6.	Solar Inverter	0.15
7.	Battery	0.6
8.	Installation	0.25
9.	Tax Burden	0.6
Total Amount		12

OPTION-2

If the product stored on rent basis as well as by own trading

A. Product stored on rent basis (say 5MT)
   Collected revenue will be = 5,000*1.8*20 = 1.8 lakhs/- per year

B. Product will stored by own trading – 5MT (say 15 rotation in a year)
   Average profit of mixed vegetable = Rs. 2.5/-
   Collected revenue will be = 5,000*2.5*15 = 2.5 lakhs per year
   Total revenue will be = 4.3 lakhs per year
   The NPV and IRR analysis of Model-3 (Option-2) is given in Table 6.10.

6.6 Result and Discussions

In this chapter, various output key parameters are investigated to identify the effects of various operating parameters. Solar DNI results in the number of hours available on different ranges at NISE, Gurugram. The power output from SPV system for operation of VDF compressor as well as thermal energy load, electricity supply to the grid through net metering, capital cost analysis, payback period, and reliability of the system have been carried out in this chapter. The performance analysis of solar radiation for Gurugram region has been investigated to find out the maximum numbers of hours available for month-wise global horizontal irradiance and direct normal irradiance.

Table 6.8 NPV and IRR analysis of Model-2.

Year	Net saving	O & M cost	Bank payment	Net expenditure	Net cash flow	Cumulative net cash flow	NPV
1.	580,000	0	152,000	152,000	428,000	428,000	−772,000
2.	**591,600**	**18,000**	**147,200**	**165,200**	**426,400**	**854,400**	**−345,600**
3.	**603,432**	**19,800**	**142,400**	**162,200**	**441,232**	**1,295,632**	**95,632**
4.	615,500	21,780	137,600	159,380	456,120	1,751,752	551,752
5.	627,810	23,958	132,800	156,758	471,052	2,222,804	1,022,804
6.	640,366	26,353	128,000	154,353	486,013	2,708,817	1,508,817
7.	653,174	28,989	123,200	152,189	500,985	3,209,802	2,009,802
8.	666,237	31,888	118,400	150,288	515,949	3,725,751	2,525,751
9.	679,562	35,076	113,600	148,676	530,886	4,256,637	3,056,637
10.	693,153	38,584	108,800	147,384	545,769	4,802,406	3,602,406
11.	707,016	42,443	104,000	146,443	560,573	5,362,979	4,162,979
12.	721,157	46,687	992,00	145,887	575,270	5,938,249	4,738,249
13.	735,580	51,356	944,00	145,756	589,824	6,528,073	5,328,073
14.	750,291	56,491	896,00	146,091	604,200	7,132,273	5,932,273
15.	765,297	62,140	848,00	146,940	618,357	7,750,630	6,550,630

Table 6.9 NPV and IRR analysis of Model-3 (Option-1).

Year	**Net saving**	**O & M cost**	**Bank payment**	**Net expenditure**	**Net cash flow**	**Cumulative net cash flow**	**NPV**
1.	360,000	0	152,000	152,000	208,000	208,000	–992,000
2.	367,200	18,000	147,200	165,200	202,000	410,000	–790,000
3.	374,544	19,800	142,400	162,200	212,344	622,344	–577,656
4.	382,034	21,780	137,600	159,380	222,654	844,998	–355,002
5.	**389,675**	**23,958**	**132,800**	**156,758**	**232,917**	**1,077,915**	**–122,085**
6.	**397,469**	**26,353**	**128,000**	**154,353**	**243,116**	**1,321,031**	**121,031**
7.	405,418	28,989	123,200	152,189	253,229	1,574,260	374,260
8.	413,526	31,888	118,400	150,288	263,238	1,837,498	637,498
9.	421,797	35,076	113,600	148,676	273,121	2,110,619	910,619
10.	430,233	38,584	108,800	147,384	282,849	2,393,468	1,193,468
11.	438,837	42,443	104,000	146,443	292,394	2,685,862	1,485,862
12.	447,614	46,687	99,200	145,887	301,727	2,987,589	1,787,589
13.	456,567	51,356	94,400	145,756	310,811	3,298,400	2,098,400
14.	465,698	56,491	89,600	146,091	319,607	3,618,007	2,418,007
15.	475,012	62,140	84,800	146,940	328,072	3,946,079	2,746,079

Table 6.10 NPV and IRR analysis of Model-3 (Option-2).

Year	Net saving	O & M cost	Bank payment	Net expenditure	Net cash flow	Cumulative net cash flow	NPV
1.	430,000	0	152,000	152,000	278,000	278,000	−922,000
2.	438,600	18,000	147,200	165,200	273,400	551,400	−648,600
3.	447,372	19,800	142,400	162,200	285,172	836,572	−363,428
4.	**456,319**	**21,780**	**137,600**	**159,380**	**296,939**	**1,133,511**	**−66,489**
5.	**465,445**	**23,958**	**132,800**	**156,758**	**308,687**	**1,442,198**	**242,198**
6.	474,754	26,353	128,000	154,353	320,401	1,762,599	562,599
7.	484,249	28,989	123,200	152,189	332,060	2,094,659	894,659
8.	493,934	31,888	118,400	150,288	343,646	2,438,305	1,238,305
9.	503,813	35,076	113,600	148,676	355,137	2,793,442	1,593,442
10.	513,889	38,584	108,800	147,384	366,505	3,159,947	1,959,947
11.	524,167	42,443	104,000	146,443	377,724	3,537,671	2,337,671
12.	534,650	46,687	99,200	145,887	388,763	3,926,434	2,726,434
13.	545,343	51,356	94,400	145,756	399,587	4,326,021	3,126,021
14.	556,250	56,491	89,600	146,091	410,159	4,736,180	3,536,180
15.	567,375	62,140	84,800	146,940	420,435	5,156,615	3,956,615

Table 6.11 IRR, ROI, and payback period of different business models.

Parameter	Model-1	Model-2	Model-3	
			Option-1	Option-2
IRR (%)	30	38	18	25
ROI (%)	29.03	35.67	17.34	23.17
Payback Period (Year)	3.5	2.9	5.8	4.4

The ground measured data have been recorded from solar radiation resource assessment station installed at NISE. It is observed that the availability of global horizontal irradiance is much higher as compared to direct normal irradiance, and this site is suitable for SPV-based cold storage as compared to solar thermal cold storage. It is also seen that around 2,700 numbers of hours available at global horizontal radiation of 300 W/m^2, in which data is more suitable and taken for designing of SPV cold storage using thermal energy storage as shown in Figure 6.6. The IRR, ROI, and payback period of different business models for the proposed system is given in Table 6.11.

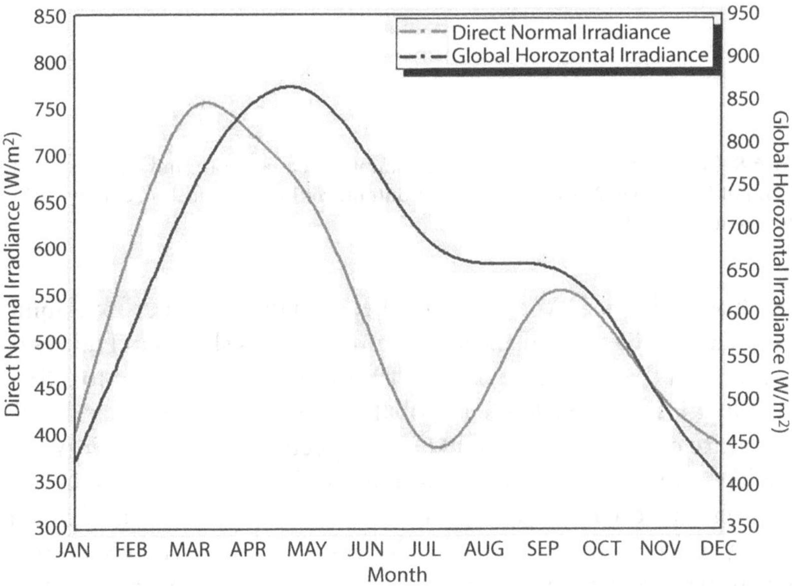

Figure 6.6 Monthly average direct normal irradiance and global horizontal irradiance at the located project site, NISE.

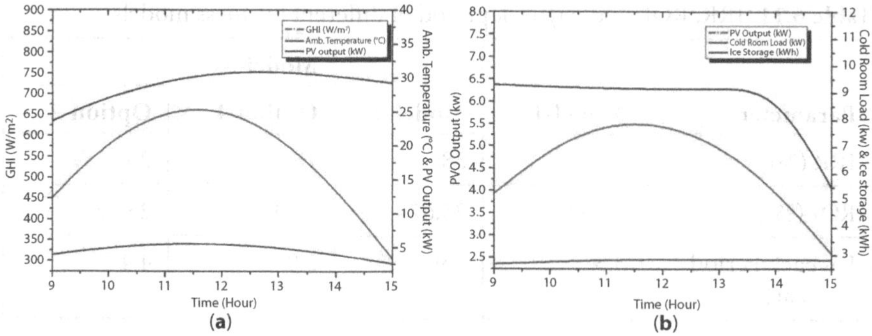

Figure 6.7 (a) Global horizontal irradiance, ambient temperature, and PV power output of month of November 2018 at NISE. (b) PV power output, cold room load, and energy store in ice storage month of November, 2018.

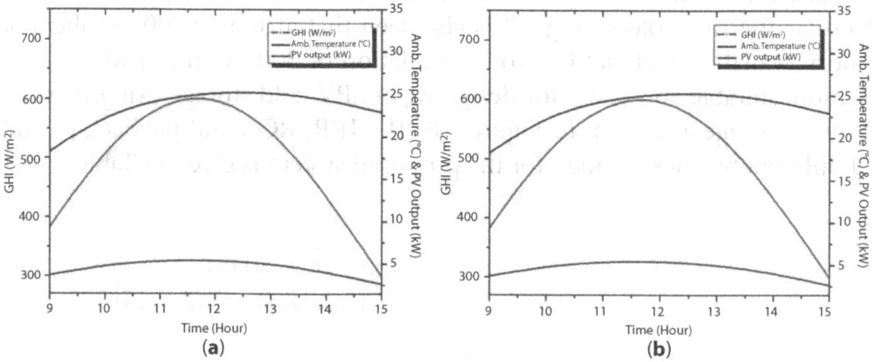

Figure 6.8 (a) Global horizontal irradiance, ambient temperature, and PV power output of month of December, 2018. (b) PV power output, cold room load, and energy store in ice storage month of December, 2018.

The hourly basis SPV output has been analyzed as per availability of global solar irradiance and ambient temperature as shown in Figures 6.7 to 6.10. It is seen that in Figure 6.7, the PV output was achieved to 5.515 kW and cooling load of 2.82 kWh at global horizontal irradiance and ambient temperature of 661 W/m^2 and 28.6°C, respectively. The ice energy storage is 9.18 kWh for continuous operation in the cloudy days and night cycle. It is also seen that the remaining power have been supplied to grid as per net metering concept, due to utilized of VFD compressor. The PV output was achieved to 5.499 kW and cooling load 2.71 kWh at global horizontal irradiance and ambient temperature of 608 W/m^2 and ambient temperature of 25.3°C, respectively, as shown in Figure 6.8.

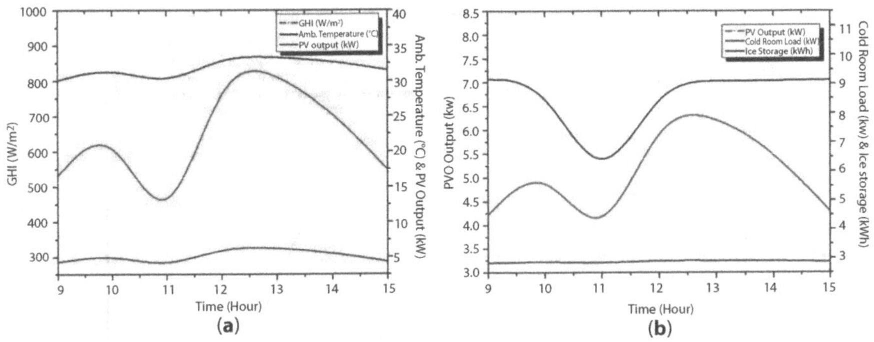

Figure 6.9 (a) Global horizontal irradiance, ambient temperature, and PV power output of month of March, 2019. (b) PV power output, cold room load, and energy store in ice storage month of March, 2019.

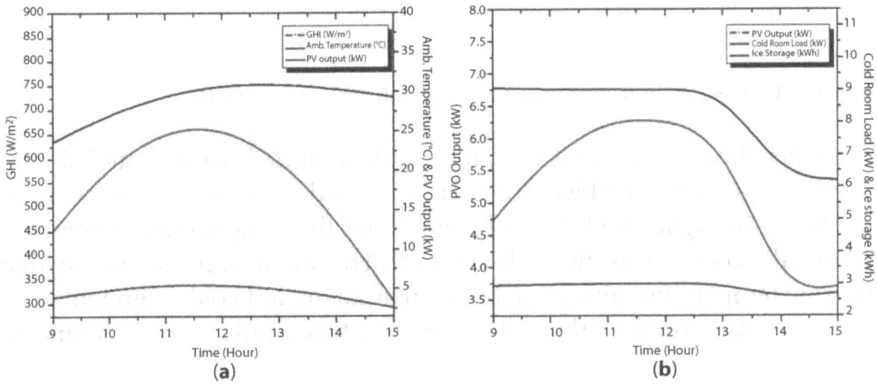

Figure 6.10 (a) Global horizontal irradiance, ambient temperature, and PV power output of month of April, 2019. (b) PV power output, cold room load, and energy store in ice storage month of April, 2019.

The maximum PV output was achieved to 6.456 kW at global horizontal irradiance and ambient temperature of 859 W/m^2 and 33.6°C, respectively, in the month of March. The ice energy storage is 9.08 kWh to supply the cooling load in non-availability of sunshine hours as shown in Figure 6.9.

In the month of April, it is investigated that the cold room load was maximum up to 3 kWh, due to increase of ambient temperature to 39.6°C. The test was carried out at no-load condition as shown in Figure 6.10. It is also seen that in the VFD compressor, the output varies as per variation of radiation and met the cooling requirement as per demand. The ice energy storage was utilized to maintain the temperature inside the cold room in cloudy and night cycle continuously for a period of 17 hours.

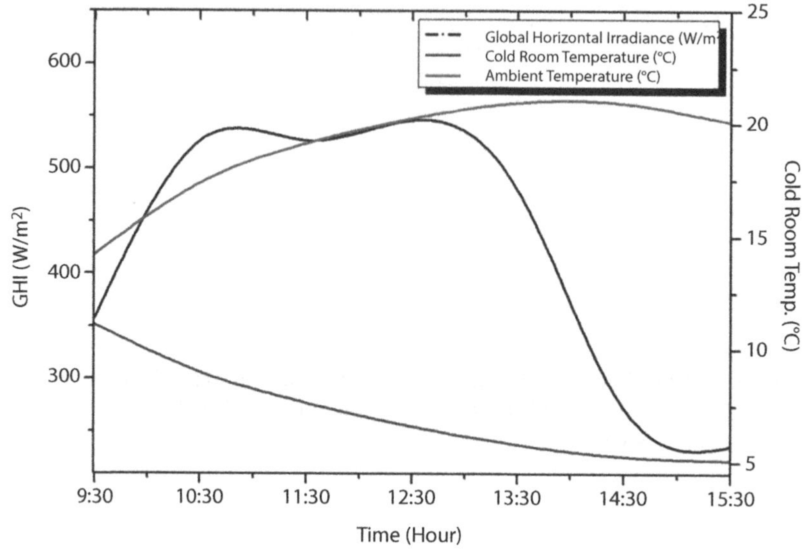

Figure 6.11 Cold room temperature variation with ambient and GHI.

As per the experimental data, the graph is plotted between global horizontal irradiance, ambient temperature, and cold room temperature as shown in Figure 6.11. It is observed that the temperature of the cold room falls very fast for first 1 hour, and after, that it decreases slowly due to large temperature difference between ambient and cold room temperature. It is also observed that cold room will take minimum 6–7 hours for

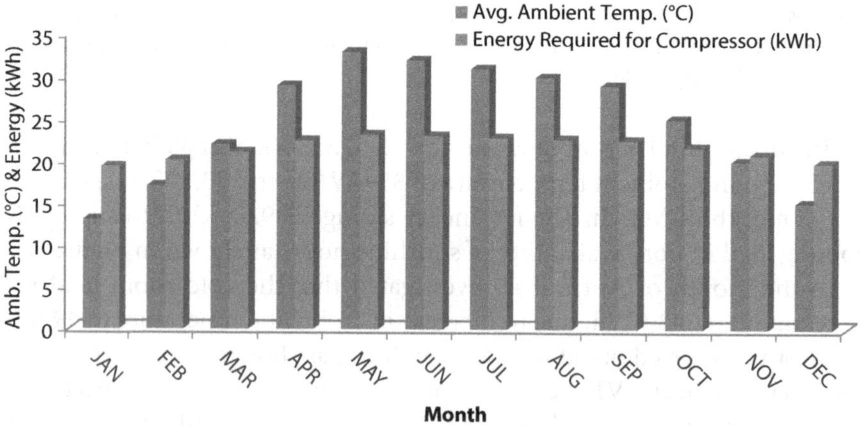

Figure 6.12 Energy required for compressor with ambient temperature on different month of a year.

maintaining the desired temperature of 4°C–5°C in no-load condition and 14–15 hours in load condition.

The ambient temperature will help to calculate the load requirement in cold storage system, and based on the load requirement pattern for different months, the actual energy required for compressor has been calculated. Figure 6.12 represents the energy requirement for compressor on monthly basis. It can be seen that the load requirement is very high in summer months (i.e., April to July) and less during winter months (i.e., November to March).

An analysis has been carried out to compare the energy generated from 7.5 kWp solar system with minimum energy required to run compressor for a VFD system. For this analysis, a solar system output of 1.2 kWe were considered as minimum energy required running compressor as VFD-based controller run from minimum of 30 Hz to a maximum of up to 75Hz. The total numbers of hours in individual months were calculated against actual energy required and the graph is plotted. From Figure 6.13. It is seen that during the month of July, the energy generated is lesser than energy required which cane managed by TES or from grid (condition to TES has no backup), whereas during other months, the energy generated is always surplus than the energy required, and hence, it can be assumed this system does not required backup energy sources for almost 11 months in a year in sunshine hour.

It is also observed that as compressor runs with the VFD controller which can take input of 1.2 kW to maximum up to rated input of 3 kW, the running time of compressor is increased to an average of 8 hours compared to constant frequency system.

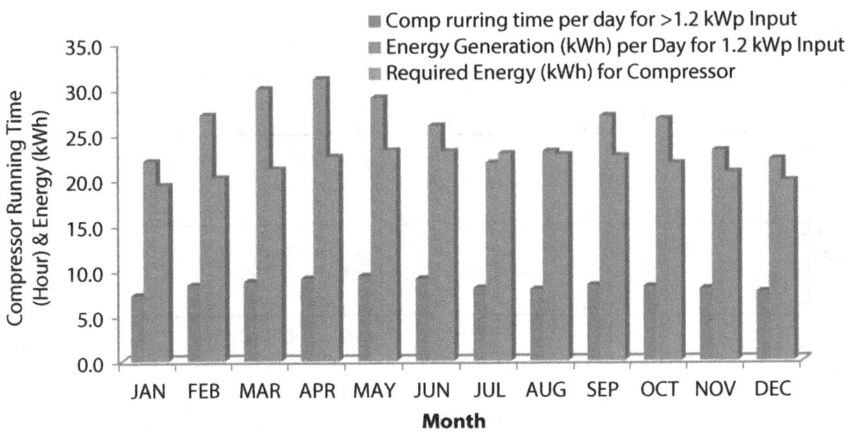

Figure 6.13 Energy generation and compressor running hours of 7.5 kW SPV plant.

The results presented in this section are based on the application of the method described earlier (VFD controller) to the system with a maximum utilization of power of up to 3 kW under normal insolation and temperature conditions. The results were obtained using computer simulation with direct current power analyzer for various insolation and temperature conditions. The proposed system showed a very satisfactory improvement compared with the system having constant frequency.

Figure 6.14 illustrates the improvement in system efficiency with different V/F ratio under different load condition. The efficiency, energy output, and power factor of the system with variable frequency, i.e., 30 Hz to 75 Hz, are able to maintain their optimum values for all simulated insolation conditions. In other words, the match between the load and source is achieved.

The new system gains more in efficiency in lower input power range since the efficiency of the system with constant frequency is affected more severely by low insolation levels than by high ones. Figure 6.14 shows the comparison of the RMS values of output voltage and current at different input power levels for the compressor load. With variable frequency control, both voltage and current tend to be less changed over whole range of the input power, resulting in a more reasonable flux level in the motor.

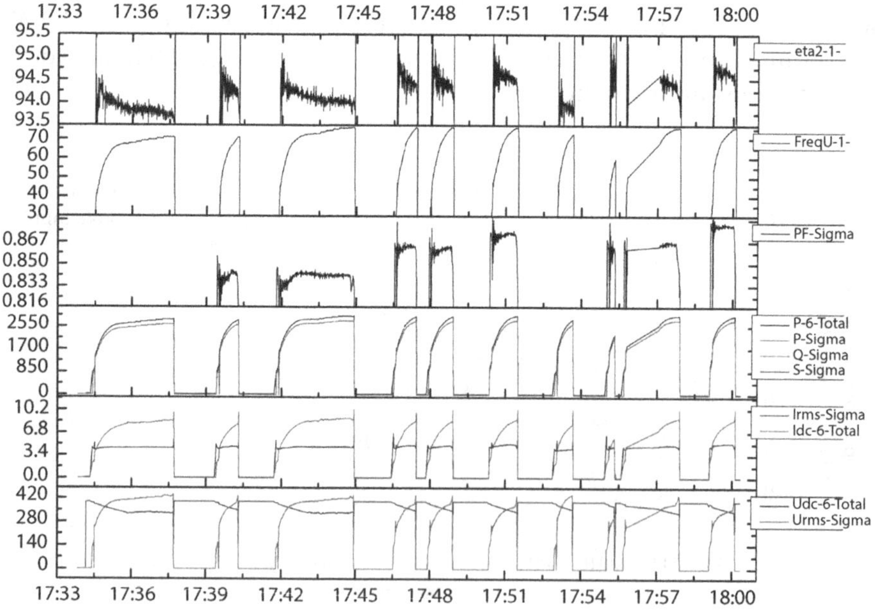

Figure 6.14 Different parameters of solar VFD controller at different V/F ratio.

The installation cost of the solar photovoltaic system and solar thermal system with the weight of component cost is shown in Tables 6.3 and 6.4. It can be seen that the thermal system has the additional costs of 5 lakhs, which is 1.4 times higher than the PV system, and also it reveals that SPV/solar collectors comprise the highest proportion of the total cost of the system, which account for 32% and 28% of the total investment, respectively.

The analysis of NPV is done on different business models, which shows (Figure 6.15) that the net gain of the project is always greater than the net expenditure value, which tells that the project is possible and a highly profitable. It is observed that the total electricity yield of the PV system is 200 MWh within 15 years and cumulative cost saving from the conventional electricity is around 12 lakhs until 15 years. It is also seen that the average IRR, ROI, and payback period of the proposed project is 27.75%, 26.3%, and 4.2, respectively, for the project life of 15 years.

At the end of research, the very crucial part is to assess the system reliability. This analysis provides the assessment of unfavorable condition in which backup for meeting actual requirement can be designed. In Figure 6.16, it can be observed that during the month of July and August, the solar energy is very less to meet actual energy demand for non-VFD system and in the month of July for a VFD system. Based on this analysis, the reliability of a VFD-based cold storage system is higher than a non-VFD–based cold storage which is 95% and 88%, respectively.

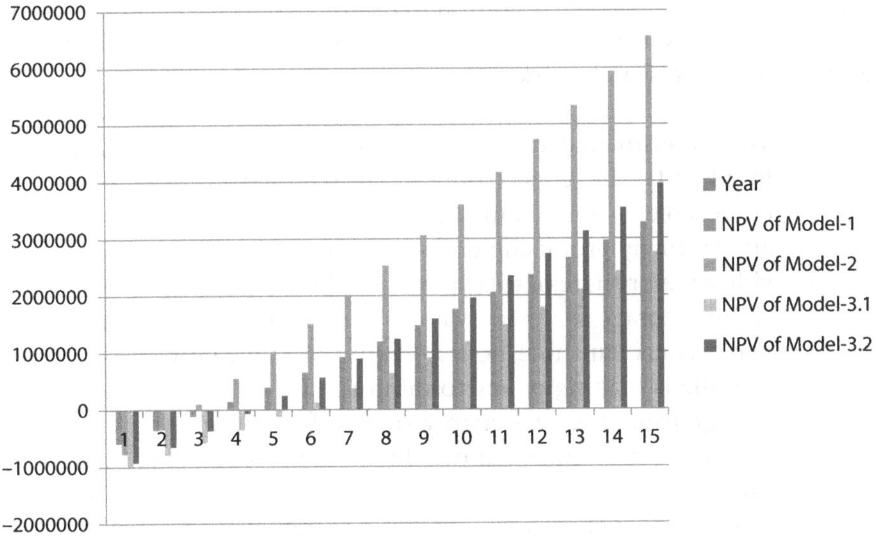

Figure 6.15 NPV of different business models.

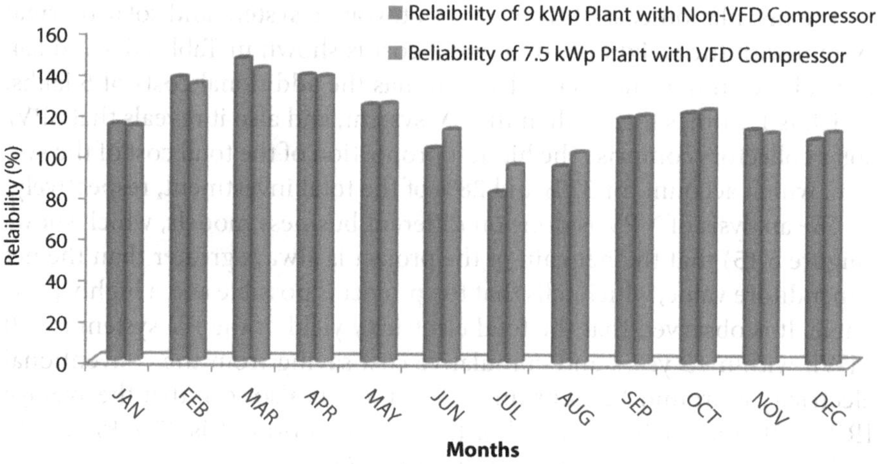

Figure 6.16 Reliability of non-VFD and VFD-based SPV cold storage.

6.7 Conclusions

In this chapter, the overall conclusions of the main findings and results have been obtained in the above research work. SPV-based cold storage cum thermal energy storage offers an opportunity to increase further efficiency through proper power management of different processes or end products as compared to stand-alone system. In the system, VFD compressor and ice thermal storage is utilized to produce cold energy for preservation of vegetables in remote areas. The following concluding remarks are drawn from the research work:

- The experimental development of the system is analyzed to identify the effects of various parameters. The SPV-based cold storage cum thermal energy storage system is an extremely promising energy system and is likely to provide a major share of solar power cold storage using thermal energy storage for preservation of vegetables and fruits as per various sunshine hours. Taking in view, Government of India is making policy to minimize the cost of SPV cold storage using thermal energy storage for creating the market. Apart from financial benefits to the promoters of these projects, the society will also be benefited due to intangible advantages such as employment generation in rural areas and environmental benefits.

- The system has achieved a maximum COP of 3.5 at various decision variables of condenser, i.e., evaporator pressure and temperature of 9.14 bars at 40°C and 1.42 bars at −5°C, respectively, of the system.
- The capital cost of the system capacity of 10 MT and thermal energy storage of 1,000 liters is Rs. 12 lakhs, which is lower than the stand-alone solar thermal cooling system, and average payback period is 4 years.

Acknowledgements

The research has been carried out at the National Institute of Solar Energy, an autonomous institute under Ministry of New and Renewable Energy, Government of India.

Abbreviations

COP	Coefficient of Performance
CAGR	Compound Annual Growth Rate
DNI	Direct Normal Irradiance
GHI	Global Horizontal Irradiance
GW	Giga Watt
IRR	Internal Rate of Return
KWh	Kilo Watt Hour
KW	Kilo Watt
Kcal	Kilo Calorie
MNRE	Ministry of New and Renewable Energy
NISE	National Institute of Solar Energy
MT	Metric Tonne
MPP	Maximum Power Point
MPPT	Maximum Power Point Tracking
MWh	Megawatt hour
NPV	Net present value
PCM	Phase Change Material
PV	Photovoltaic
ROI	Return on Investment
SCM	Supply Chain Management
SPV	Solar Photovoltaic

TES	Thermal Energy Storage
TR	Tones of Refrigeration
TX	Thermal Expansion
USD	United States Dollar
Voc	Open Circuit Voltage
VFD	Variable Frequency Drive
VARS	Vapor Absorption Refrigeration System
VCRS	Vapor Compression Refrigeration System
Wp	Watt Peak

References

1. National Centre for Cold Chain Development (NCCCD), (http://www.nccd.gov.in), 23.12.2020.
2. Cold storage capacity report, Global Cold chain Alliance, July 2018, (https://www.gcca.org/sites/default/files/2018%20GCCA%20Cold%20Storage%20Capacity%20Report%20final.pdf).
3. Om Prakash, Anil Kumar. Solar Drying Technology: Concept, design, testing, modelling, economics and environment. Green Energy and Technology, 3–33, 2017.
4. Business Standard Private Ltd., Economy & Policy Report, March 23, 2019. [https://www.business-standard.com/article/economy-policy/cold-storage-capacity-expands-rs-21-000-crore-investment-lined-up-by-2023-119032200623_1.html#:~:text=The%20current%20cold%20storage%20capacity,is%20multi%2Dcommodity%20cold%20storage.] 25.11.2020.
5. Sahoo, U., Kumar, R., Pant, P.C., Chaudhary, R., Resource assessment for hybrid Solar-Biomass power plant and its thermodynamic evaluation in India. Solar Energy, 139, 47–57, 2016.
6. Sahoo, U., Kumar, R., Pant, P.C., Chaudhary, R., Scope and sustainability of hybrid solar-biomass power plant with cooling, desalination in polygeneration process in India. Renew. Sust. Energy Rev., 51, 304–316, 2015.
7. S. Kumar, V.Yadav, U.Sahoo, S.K.Singh. Experimental investigation of 16 square meter Scheffler concentrator system and its performance assessments for various regions of India. Thermal Science and Engineering Progress, 10: 103–111, 2019.
8. U. Sahoo, S.K. Singh, R.Kumar, P.C.Pant, I.Barbate. Experimental study of an inclined Flat Plate type Solar Water Distillation System. Renewable: wind, water and solar, 3:2–5, 2016.
9. U. Sahoo, S.K. Singh, R.Kumar, P. Kumar. Mathematical Modelling of Portable Solar Water Distillation System. Journal of Technology Innovations in Renewable Energy, 4: 91–95, 2015.

10. U. Sahoo, S.K. Singh, J.P. Singh, H.Sharma, P.Kumar., Design & Performance Analysis of Portable Solar Water Heating System for Rural Areas: Himalayan Regions, India, International Journal of Renewable Energy Research, 4:743–748, 2014.
11. Sahoo U (2018). A polygeneration process concept for hybrid solar and biomass power plant: simulation, modelling, and optimization. Wiley.
12. K.Sahoo, B. Bandhyopadhyay, S. Mukhopadhyay, U. Sahoo, T. S. Kumar, V. Yadav, Y. Singh., Cold Storage with Backup Thermal Energy Storage System. Advances in Renewable Energy: Progress in Solar Energy Technologies and Applications, Wiley & Scrivener Publishing 2019, pg. 181–232.
13. U Sahoo, SK Singh, R Kumar, PC Pant, P Saxena. Evaluation of Solar Thermal Technologies and Applications in India, Advances in Energy Research, 21, 141–189, 2015.
14. U. Sahoo, Modelling and Optimization of A Polygeneration Hybrid Solar and Biomass System for Power, Cooling and Desalination, PhD Thesis, Delhi Technological University, India, Report no. TD-3098, viewed 11th May 2019, http://dspace.dtu.ac.in:8080/jspui/handle/repository/16092>.
15. Sahoo, U., Kumar, R., Singh, S.K., Tripathi, A.K., Energy, Exergy, Economic Analysis and Optimization of Polygeneration Hybrid Solar-Biomass System. Appl. Therm. Eng., 145, 685–692, 2018.
16. Sahoo, U., Kumar, R., Pant, P.C., Chaudhary, R., Development of an innovative polygeneration process in hybrid solar-biomass system for combined power, cooling and desalination. Appl. Therm. Eng., 120, 560–567, 2017.
17. P.S. Chauhan, A. Kumar, C. Nuntadusit, J. Banout. Thermal modeling and drying kinetics of bitter gourd flakes drying in modified greenhouse dryer. Renewable Energy 118, 799–813, 2018.
18. R. Dejchanchaiwong, Y. Tirawanichakul, S. Tirawanichakul, A. Kumar, P. Tekasakul. Conjugate heat and mass transfer modeling of a new rubber smoking room and experimental validation. Applied Thermal Engineering 112, 761–770, 2017.
19. Kart, M.C.O., Demircan, V., An economic comparison of conventional and modern cold storage facilities in Turkey, 10 (1), 118–130, 2014. (http://www.custoseagronegocioonline.com.br/numero1v10/Storage.pdf)

7

Estimation of Fault Voltages in Renewable Energy–Based Microgrid

Golla Anand, Chinmoy Basak, Rishabh Anand, Sourav Sahoo and Prof. Sarita Nanda*

Kalinga Institute of Industrial Technology (KIIT Deemed to be University) Bhubaneswar, Odisha, India

Abstract

Renewable sources of energy nowadays are the leading technologies in almost every field of application to protect the environment from global warming caused by consumption and burning of fossil fuels in large quantities. This also leads to the rapid exhaustion of fossil fuels, and the day is not far off when there will be no fossil fuels left. So in order to shift the major part of our day-to-day loads from fossil fuels to renewable sources we need to use renewable sources efficiently and extract the best possible energy available with them. But using renewable energy sources has its own challenges. They mostly depend on natural sources, which are not always constant and vary over time, causing variation in power generation. Let's say that wind energy, the power generated by wind mills, is directly proportional to the wind speed. As we know, wind speed is never constant. It keeps on changing. The same is true for solar energy. The amount of power generated is directly proportional to the amount and intensity of sunlight on the solar modules. This variation in power generation results in various kinds of disturbances, such as variation in power output, variation in frequency, voltage or current harmonics and sag and swell fluctuation and flickering in the voltage resulting in peaks and troughs. Thus, fault voltage/current estimation with high estimation accuracy and low computational complexity is necessary for effective real-time monitoring and control of microgrid interfacing to the main grid. Most of the methods used are Real-Time Phasor Estimation and Discrete Fourier Transform (DFT). This chapter presents a simple approach and its implementation for the estimation of fundamental, harmonic phasor and frequency of time varying current and voltage signals using a modified Taylor-based RLS learning algorithm.

**Corresponding author*: snandafet@kiit.ac.in

Umakanta Sahoo (ed.) Hybrid Renewable Energy Systems, (169–182) © 2021 Scrivener Publishing LLC

Keywords: Taylor-recursive least square (rls), phasor estimation, renewable energy, smart grid, microgrid, phasor measurement unit (PMU)

7.1 Introduction

India is one of the world leaders in terms of the production of energy from renewable sources. As per a report published on 31st March 2020, 35.86% of India's total power generation capacity is from renewable sources, including about 21.22% of total utility electricity in the country. In the Paris Agreement India has committed to Intended Nationally Determined Contributions to achieve a target of 40% of its total electricity generation from renewable and non-fossil fuel sources by 2030. A high demand for renewable and non-fossil fuels energy is given emphasis for various reasons. The future shortage of fossil fuels, environmental concerns, and advancement in smart grid technologies, all stimulate an increasing penetration of renewable energy. According to its 2027 blueprint, India plans to achieve a goal of 275 GW from renewable energy, 72 GW of hydroelectricity, 15 GW of nuclear energy and nearly 100 GW from "other zero emission" sources. To control the inflow and outflow to such huge unbalanced energy generation, it is important to implement smart grid technologies to the existing systems and upgrade them to sustain every situation in the future.

A Smart Grid is a type of electricity network or grid which has an efficient cost of production and can process the behavior and actions of all users, i.e., the amount of generation and amount of consumption connected to it – generators, consumers and those that do both – in order to ensure an efficient, sustainable power system with minimal amount of losses and high levels of quality and reliability of supply and safety. The smart grid Electricity network is shown in Figure 7.1. It deals with improved production and functions together with good monitoring control, communication, and self-healing technologies so that:

- The connection and operation of generators of all sizes and technologies will have better facility.
- Consumers can play a part in altering the operation of the system.
- Relevant information can be provided to the consumers for how they will use their supply.
- Environmental impact of the whole electricity supply system can be reduced significantly.

- The existing high levels of system reliability, quality and reliability of supply can be maintained or even improved.
- The existing services can be efficiently maintained and improved.

A smart grid is a combination of micro- and minigrids with fine branch and supply systems control. Microgrids are small grid networks of a power supply system that are designed to provide power supply to a small area or community. A microgrid basically includes local loads and local generator. It comprises numerous small power generating sources which make it highly flexible and efficient in meeting the required demand. The microgrid is connected to both the local generating units and utility grid, thus preventing power outage during the loss of generation from local power generators. Also the excess power produced can be sold to the utility grid where utility is an electric power distribution for commercial purpose which uses wind turbine, water turbine, diesel generator, etc.

The size of a microgrid depends on the region which it supplies. This may range from the size of a housing estate to a size of municipal regions.

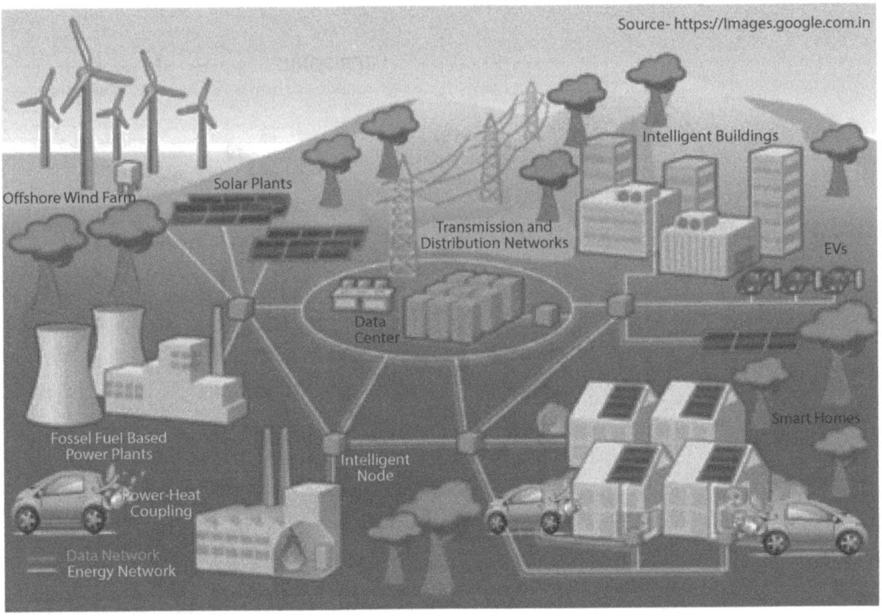

Figure 7.1 Smart grid network.

The desirability of a microgrid is due to following reasons:

1. It can play an important role on the energy crisis.
2. Transmission losses gets minimized.
3. Provides high-quality, reliable and uninterrupted energy supply to critical loads.

A common microgrid layout is shown in Figure 7.2.

While there are centralized renewable generation facilities like solar parks or wind farms, a different and appealing approach is to use renewable energy locally. It is important to consider the intermittency of renewable generation [1, 2] and its varying geographical and temporal availability to avoid blackouts and make the grid more robust. The weather and its impact on renewable energy generation makes it important to estimate the energy signals and disconnect unbalanced generations or loads from the grid to avoid blackouts and imbalances in the grid. This can be done with the use of **Phasor Measurement Unit (PMU)** [1, 3–5]. This is a device used to estimate the magnitude and phase angle of an electrical phasor quantity (such as voltage or current) in the electricity grid using a common time source for synchronization.

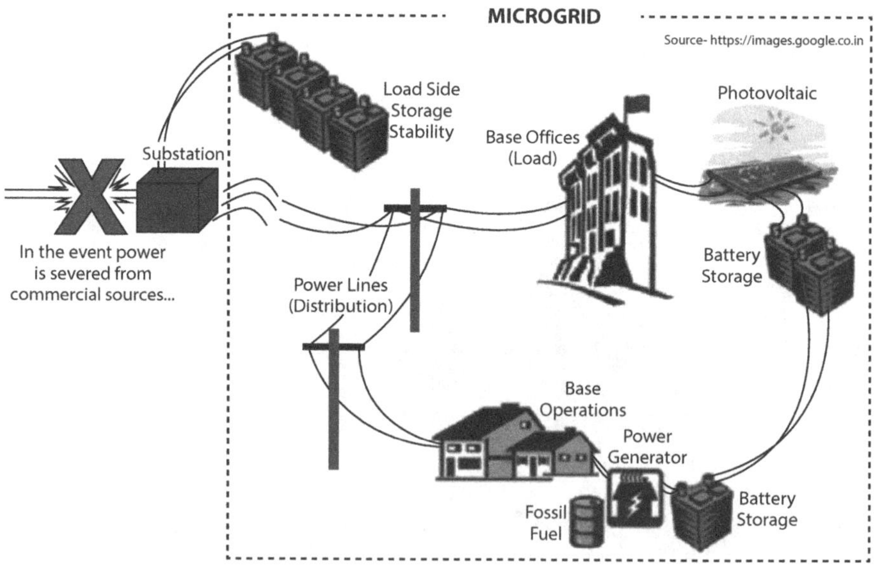

Figure 7.2 Microgrid network.

Different methods for phasor estimation can be found in the literature like Zero Crossing, Discrete Fourier Transform (DFT), Kalman Filter [6], Demodulation, etc. Here, we have used a new approach to it, i.e., Taylor Series based Recursive Least Square (RLS) Algorithm [1, 2].

7.2 Problem Formulation

7.2.1 Taylor Series Based Voltage Signal Formulation

A non-stationary voltage signal with amplitude and phase components can be represented as:

$$y(n) = A(n) \cdot \sin\left[2\pi f(n) n \, dt + \varphi(n)\right] \tag{7.1}$$

where A(n) and φ(n) are the non-stationary amplitude and phasor at the nth sampling instant of the signal, respectively, dt is given as the sampling time and fc(n) is given as the fundamental frequency component of the given signal. Now θ can be defined as the angular component of the signal which can be given by $\theta(n) = (2\pi fc(n) n \, dt + \varphi(n))$.

Now the trigonometric expansion of (7.1) can be presented as

$$\begin{aligned} y(n) = {} & A(n) \cdot \cos \varphi(n) \cdot \sin(2\pi fc(n) n \, dt) \\ & + A(n) \cdot \sin \varphi(n) \cdot \cos(2\pi fc(n) n \, dt) \end{aligned} \tag{7.2}$$

Again (7.2) can be rewritten as

$$y(n) = Xc(n) \cdot \sin(2\pi fc(n) n \, dt) + Xs(n) \cdot \cos(2\pi fc(n) n \, dt) \tag{7.3}$$

Where

$$Xc(n) = A(n)\cos \varphi(n) \text{ and } Xs(n) = A(n)\sin \varphi(n) \tag{7.4}$$

Xc(n) and Xs(n) are defined as the dynamic envelope of the non-stationary sinusoidal signal. Thus these coefficients are expanded using Taylor series [7, 8] till second order, as given below

$$Xc \cong to_0 + to_1 n \, dt + f \, to_2 n \, dt^2 \tag{7.5}$$

$$Xs \cong ti_0 + ti_1 n \, dt + ti_2 n \, dt^2 \tag{7.6}$$

Where the Taylor coefficients till first order are given by

$$\mathrm{to}_0 = Xc(0) = A(0)\cos\phi(0) \tag{7.7}$$

$$\mathrm{to}_1 = \frac{d}{dt}[Xc(0)] = \frac{d}{dt}A(0)cos\phi(0) - A(0)\frac{d}{dt}sin\phi(0) \tag{7.8}$$

$$\mathrm{ti}_0 = Xs(0) = A(0)\sin\phi(0) \tag{7.9}$$

$$\mathrm{ti}_1 = \frac{d}{dt}[Xs(0)] = \frac{d}{dt}A(0)sin\phi(0) + A(0)\frac{d}{dt}cos\phi(0) \tag{7.10}$$

In (7.8) and (7.10), for the purpose of determining the time varying frequency, the amplitude component $A(0)$ is assumed to be constant and its derivative is zero. Hence, substituting (7.7) and (7.9) into (7.8), and (7.10) and by neglecting $d/dt[A(0)]$, the derivative of the phase angle is obtained as

$$\frac{d}{dt}[\phi(0)] = \frac{to_0 ti_1 - to_1 ti_0}{to_0^2 + ti_0^2} \tag{7.11}$$

The phasor estimation of the fundamental component of a non-stationary sinusoidal signal can also be given as in (7.12) and (7.13) which can be expanded for harmonic estimation also

$$A(n) = \sqrt{to_0^2 + ti_0^2} \tag{7.12}$$

$$\phi(n) = \arctan\left(\frac{ti_0}{to_0}\right) \tag{7.13}$$

where $fc(n)$ is the estimate of the frequency. Now by using the second-order Taylor series model, the non-stationary sinusoidal signal in (7.2) can be described as

$$\begin{aligned} y(n) &= (\mathrm{to}_0 + \mathrm{to}_1 n\,dt + \mathrm{to}_2 n\,dt^2) \cdot \sin(2\pi fc(n)n\,dt) \\ &\quad + (\mathrm{ti}_0 + \mathrm{ti}_1 n\,dt + \mathrm{ti}_2 n\,dt^2) \cdot \cos(2\pi fc(n)n\,dt) \end{aligned} \tag{7.14}$$

The input vector to the linear combiner is represented as $\alpha(n)$ which is sampled at a rate of $dt = 1/fs$, where fs is defined as the sampling frequency given in (7.15) and $W(n)$ is defined as the weight vector at nth sampling instant presented in (7.16) as follows:

$$\alpha(n)=\begin{bmatrix} \sin(2\pi fcndt) & ndt\sin(2\pi fcndt) & (ndt)\wedge 2\sin(2\pi fcndt) \\ \cos(2\pi fcndt) & ndt\cos(2\pi\ fcndt) & (ndt)\wedge 2\cos(2\pi fcndt) \end{bmatrix} \tag{7.15}$$

$$\hat{W}(n) = [to_0\ to_1\ to_2\ ti_0\ ti_1\ ti_2] \tag{7.16}$$

Now the adjustment of the weight vector $\hat{W}(n)$ of the linear combiner can be obtained by using a non-linear weight adaptation rule given by:

$$\hat{W}(n+1)=k(n)\hat{W}(n)+\frac{\mu(n)\varepsilon(n)\alpha^T(n)}{\alpha(n)\alpha^T(n)} \tag{7.17}$$

where $\varepsilon(n)$ is the estimated error signal given as in (7.18), $\boldsymbol{\alpha}(n)$ is the input vector, $\mu(n)$ and $k(n)$ are the learning parameter and variable factor respectively. The error signal $\varepsilon(n)$ can be calculated as

$$\varepsilon(n) = y(n) - \hat{W}(n-1)\alpha^T(n) \tag{7.18}$$

7.2.2 Recursive Least Square (RLS) Algorithm

Thus input signal vector can be expressed as

$$\alpha(n)=\begin{bmatrix} \sin(2\pi fcn\ dt) & n\ dt\sin(2\pi fcn\ dt) & (n\ dt)\wedge 2\sin(2\pi fcn\ dt) \\ \cos(2\pi fcn\ dt) & n\ dt\cos(2\pi\ fcn\ dt) & (n\ dt)\wedge 2\cos(2\pi fcn\ dt) \end{bmatrix} \tag{7.19}$$

The weight vector can be expressed as:

$$\hat{\mathrm{W}}(n) = [to_0\ to_1\ to_2\ ti_0\ ti_1\ ti_2] \tag{7.20}$$

Estimated amplitudes and phases can be evaluated from the updated weight vector obtained by using RLS algorithm [9, 10, 12] as given from Eq. (7.21) to Eq. (7.26)

$$a_{i(1)} = \sqrt{(W_{k(1)})^2 + (W_{k(2)})^2} \tag{7.21}$$

$$a_{i(3)} = \sqrt{(W_{k(3)})^2 + (W_{k(4)})^2} \tag{7.22}$$

$$a_{i(5)} = \sqrt{(W_{k(5)})^2 + (W_{k(6)})^2} \tag{7.23}$$

$$\phi_{i(1)} = \tan^{-1}\left(\frac{(W_{k(2)})}{(W_{k(1)})}\right) \tag{7.24}$$

$$\phi_{i(3)} = \tan^{-1}\left(\frac{(W_{k(4)})}{(W_{k(3)})}\right) \tag{7.25}$$

$$\phi_{i(5)} = \tan^{-1}\left(\frac{(W_{k(6)})}{(W_{k(5)})}\right) \tag{7.26}$$

7.3 Pseudo Code/Algorithm for Taylor-RLS

Step 1

The voltage signal imported from the Wind-Solar model is converted into its Taylor equivalent using Taylor series expansion.

$$\alpha(n) = \begin{bmatrix} \sin(2\pi fcn\ dt) & n\ dt \sin(2\pi fcn\ dt) & (n\ dt)\wedge 2\sin(2\pi fcn\ dt) \\ \cos(2\pi fcn\ dt) & n\ dt\cos(2\pi\ fcn\ dt) & (n\ dt)\wedge 2\cos(2\pi fcn\ dt) \end{bmatrix}$$

Step 2
The weight vector 'W' matrix and Inverse correlation matrix 'Q' are initialized below.

$$\hat{W}\,(n) = [to_0\ to_1\ to_2\ ti_0\ ti_1\ ti_2]$$

$$Q=\frac{1}{\delta}\begin{bmatrix} 1 & 0 & \ldots\ldots\ldots\ 0 \\ 0 & 1 & \ldots\ldots\ldots\ 0 \\ 0 & \ldots\ldots\ldots & \ldots\ldots\ldots\ 1 \end{bmatrix}$$

Step 3
The estimated voltage signal 'y(n)' is calculated by multiplying weight vector with the input signal.

Step 4
Then the error 'ε(n)' between the original voltage signal and the estimated voltage wave form is calculated.

Step 5
Inverse correlation matrix can be updated as given below:

$$\psi(n) = Q\,(n) \times \alpha\,(n)$$

$$inv(n+1)=\frac{1}{\beta}\left[inv(n)-\frac{\psi_n\psi_n^T}{\lambda+\psi_n^T\alpha(n)}\right]$$

Where, β = forgetting factor

Step 7
Then the weight vector 'W' is updated by using:

$$W\,(n+1) = W\,(n) + \varepsilon\,(n)\,Q\,(n+1)\,\alpha\,(n)$$

7.4 Experimental Validation

The proposed algorithm is verified for a non-stationary sinusoidal wind-solar renewable signal model using MATLAB/Simulink [15, 16] as shown

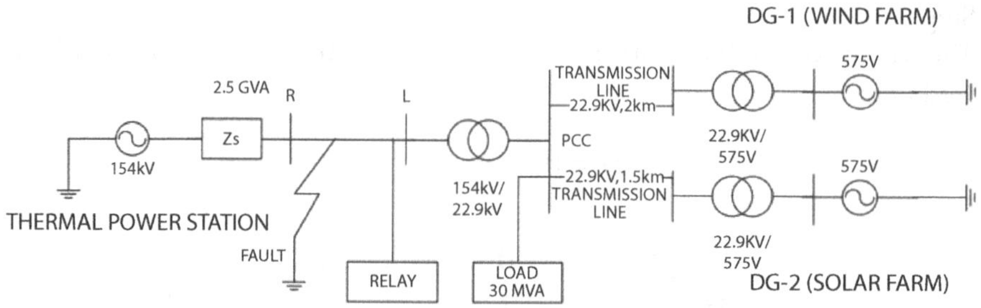

Figure 7.3 Wind-solar model integration with microgrid.

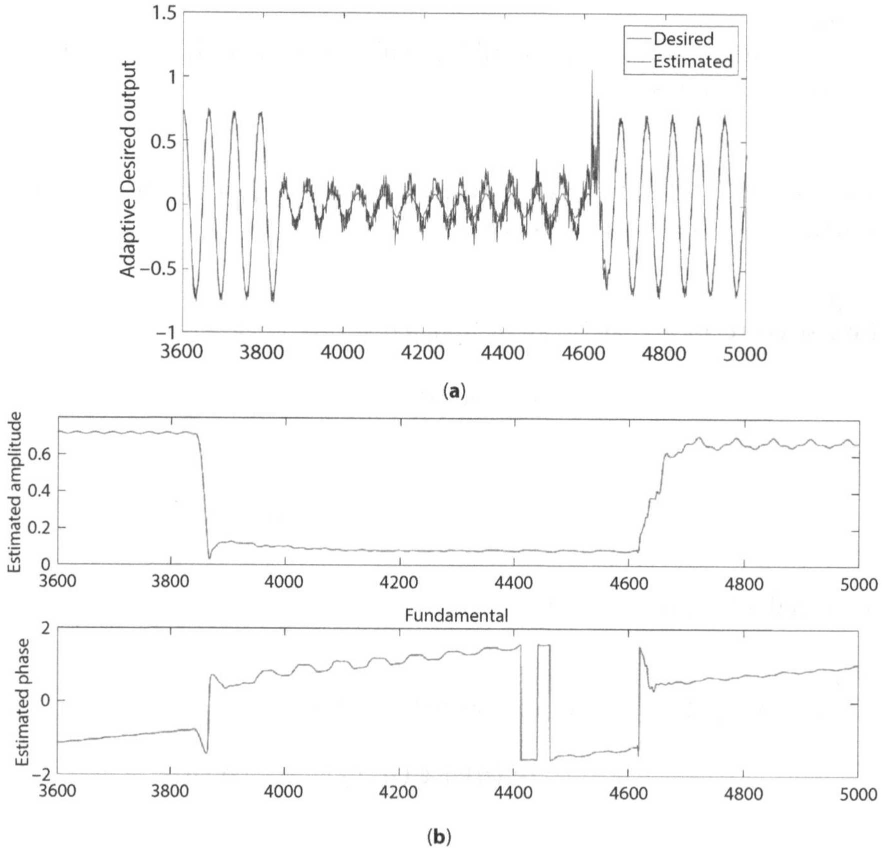

Figure 7.4 (a) Adaptive desired output with RLS algorithm for wind-solar model. (b) Estimated amplitude and phase for the wind-solar model.

in Figure 7.3, where energy is generated from the thermal plant, wind and solar farms and are synchronised with the main grid. Here the significance of PMU [2, 4, 7] and its application in smart grid comes into picture.

In this case, the fault signal is simulated by the three-phase fault generator in the MATLAB/Simulink environment where Wind-Solar farm connected to a distributed transmission line [11, 12, 13, 14] is modelled. Single line to ground fault (L-G) was simulated on the 154 KV transmission line at a distance of 16KM from the L bus during 1-1.2s interval. The phasor measurement by Taylor Based RLS algorithm is shown in Figure 7.4(a). The mean square voltage estimation errors for each phase in wind-solar model is given in Table 7.1.

In the second case, a microgrid-based smart grid 500 KV transmission line is taken into consideration and Line to Line (L-L) fault was applied to this transmission line at a distance of 100Km as shown in Figure 7.5. The three-phase fault voltage signal is shown in Figure 7.6 (a).

The adaptive desired output by Taylor Based RLS algorithm is shown in Figure 7.6(b) and the mean square voltage estimation errors for each phase in L-L fault model is given in Table 7.2.

Table 7.1 Mean square voltage estimation error.

Phase A	Phase B	Phase C
0.0014	0.0037	0.0023

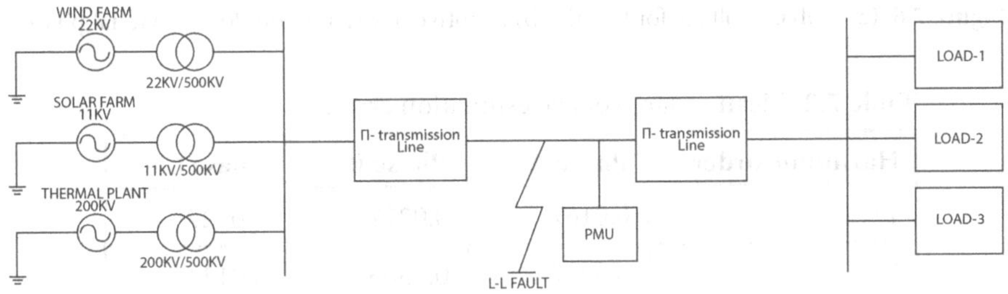

Figure 7.5 Distributed Pi-network transmission line.

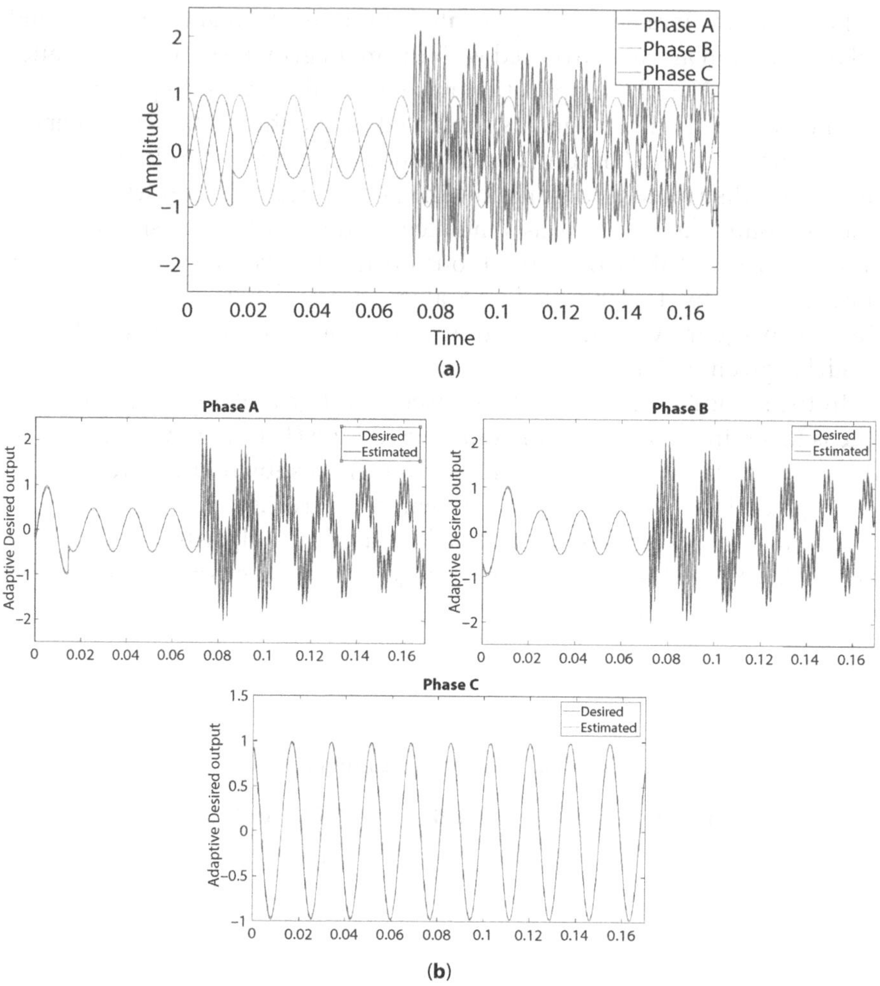

Figure 7.6 (a) 3 phase voltage for ll fault, (b) adaptive desired output for individual phase.

Table 7.2 Mean square voltage estimation error.

Harmonic order	Phase A	Phase B	Phase C
1st	0.0167	0.0272	0.0127
3rd	0.0217	0.0306	0.0134
5th	0.0340	0.0419	0.0140
11th	0.0420	0.0507	0.0159

7.5 Conclusion

A new and robust Recursive Least Square (RLS) based on Taylor Series expansion has been proposed to estimate the non-stationary amplitude and phase of the fundamental and decaying signals present in a power signal. The experimental results for the fundamental amplitude and phase estimation of non-stationary power signals mixed with harmonics and noise clearly demonstrate the accuracy of the proposed algorithm. Moreover, the errors are significantly reduced using Taylor series based RLS algorithm as compared to any other existing algorithms which are used in PMUs. With this accuracy the estimation of fault voltage and current signals in renewable energy–based microgrid will be more reliable and we will be able to detect and disconnect the faulty line from the main grid more quickly and efficiently, which is in demand today.

References

1. Frigo G, Derviškadić A, Pegoraro PA, Muscas C, Paolone M. Harmonic Phasor Measurements in Real-World PMU-Based Acquisitions. In *2019 IEEE International Instrumentation and Measurement Technology Conference (I2MTC) 2019 May 20* (pp. 1-6). IEEE.
2. Vanfretti L, Baudette M, White AD. Monitoring and control of renewable energy sources using synchronized phasor measurements. *Renewable Energy Integration* 2017 Jan 1 (pp. 419-434). Academic Press.
3. Yang P. Power System State Estimation and Renewable Energy Optimization in Smart Grids.
4. Nanda S, Dash PK. A Gauss–Newton ADALINE for dynamic phasor estimation of power signals and its FPGA implementation. *IEEE Transactions on Instrumentation and Measurement.* 2016 Nov 16;67(1):45-56.
5. Hojabri M, Dersch U, Papaemmanouil A, Bosshart P. A comprehensive survey on phasor measurement unit applications in distribution systems. *Energies.* 2019 Jan;12(23):4552.
6. Nanda S. Kalman filters for parameter estimation of nonstationary signals. *Kalman Filters-Theory for Advanced Applications* 2018 Feb 21. InTech.
7. Nanda S, Dash PK. Field programmable gate array implementation of fuzzy variable step size adaptive linear element for adaptive frequency estimation. *IET Signal Processing.* 2017 Jul 18;11(9):1083-94.
8. Petrović P, Damljanović N. Dynamic Phasors Estimation Based on Taylor-Fourier Expansion and Gram Matrix Representation. *Mathematical Problems in Engineering.* 2018 Jan 1;2018.

9. Sahoo HK, Subudhi U. Adaptive estimation of power system harmonics and decaying DC using Volterra RLS. In *2015 IEEE Power, Communication and Information Technology Conference (PCITC)* 2015 Oct 15 (pp. 43-47). IEEE.
10. Kamran M, Azam F, Khanum A. Discovering core architecture classes to assist initial program comprehension. In *Proceedings of the 2012 International Conference on Information Technology and Software Engineering 2013* (pp. 3-10). Springer, Berlin, Heidelberg.
11. Begovic MM, Djuric PM, Dunlap S, Phadke AG. Frequency tracking in power networks in the presence of harmonics. *IEEE Transactions on Power Delivery.* 1993 Apr;8(2):480-6.
12. Abdollahi A, Matinfar F. Frequency estimation: A least-squares new approach. *IEEE Transactions on Power Delivery.* 2010 Dec 13;26(2):790-8.
13. Lee DG, Kang SH, Nam SR. Modified dynamic phasor estimation algorithm for the transient signals of distributed generators. *IEEE Transactions on Smart Grid.* 2012 Dec 28;4(1):419-24.
14. Cardenas A, Guzman C, Agbossou K. Development of a FPGA based real-time power analysis and control for distributed generation interface. *IEEE Transactions on Power Systems.* 2012 Mar 6;27(3):1343-53.
15. Nanda S, Biswal M, Dash PK. Estimation of time varying signal parameters using an improved Adaline learning algorithm. *AEU-International Journal of Electronics and Communications.* 2014 Feb 1;68(2):115-29.
16. Nanda S, Dash PK, Chakravorti T, Hasan S. A quadratic polynomial signal model and fuzzy adaptive filter for frequency and parameter estimation of nonstationary power signals. *Measurement.* 2016 Jun 1;87:274-93.

8

Optimization of PV-Wind Hybrid Renewable Energy System for Health Care Buildings in Smart City

A. Karthick*, V. Kumar Chinnaiyan, J. Karpagam, V.S. Chandrika and P. Ravi Kumar

Department of Electrical and Electronics Engineering, KPR Institute of Engineering and Technology, Arasur, Coimbatore, Tamilnadu, India

Abstract

Hybrid energy resources are a promising technology for future energy sources. In this work, Solar Photovoltaic and wind energy are combined to form the Hybrid energy resources. The Hybrid Optimization Model for Electric Renewables (HOMER) computational tool is used to analyze the Hybrid energy sources and considering the health care building electrical utility as a load. The result revealed that 3 kW photovoltaic system could generate the 4,437 kWh/yr and 3 kW wind turbine 577 kWh/yr for the Chennai region (13.0827° N, 80.2707° E) Tamilnadu, India. Total energy consumption of the health care building is 4,109 kWh/yr predicted by HOMER. During a summer day, the proposed hybrid energy system produced the required power to satisfy the demand. During a windy day, the Wind turbine operated and supplied the required power for the utilization of the required energy. The generator generated 584 kWh/yr power for Chennai city. As is clear, the most elevated and least degree of solar radiation with 225 kWh/m^2/day and 150 kWh/m^2/day, happened in April and July, respectively. It is found that the net present cost of the system obtained during the tenth year from the system installed. The hybrid energy system saves around 495 kg/yr in carbon dioxide. The renewable energy fraction of the system is found to be 0.86.

Keywords: Photovoltaic, diesel generator, hybrid energy, wind energy, HOMER

**Corresponding author*: karthick.power@gmail.com

Umakanta Sahoo (ed.) Hybrid Renewable Energy Systems, (183–198) © 2021 Scrivener Publishing LLC

8.1 Introduction

In India and Africa, where the vast population is increasing, most rural residents are still disconnected from electricity supply. A disconnected grid network able to produce electricity on-site is considered a feasible alternative for these regions' power supplies. Sustainable energy sources such as biomass and solar photovoltaics (PV) systems are electrified in the villages of India [1]. Solar is the world's largest energy source (170,000 TWhr) and is easily accessible via established innovations. The global change to a safe and sustainable hybrid energy system is largely due to two reasons: the potential technological advantages of hybrid mixtures and the rapid explosion of conventional energy sources. Much of India is without electricity in the remote area. Diesel generators (DGs) are used in remote areas to satisfy required load demand. Because of cost variation in fossil fuel and maintenance of the DG's, operating costs are high. Through its various hybrid combinations, alternative energy sources such as photovoltaics and wind provide suitable options for electricity generation off-grid. Photovoltaics and wind renewable energy (PV) is seen as the primary non-conventional energy production option in an ever-growing energy crisis in terms of productivity, such as environmentally friendly, volume availability and locational reliance. The downside is that the total costs of the service cycle are raised and reduced. In the *Economic Times* report it is estimated that total RE potential for India (up to 2022) is 175 GW [2]. Two factors are largely behind the overall demand for strong and feasible half-and-half renewable energy sources; the potential technological preferences of cross-breed combinations and the rapid depletion of traditional sources of vitality. In India, a large part of the distant area doesn't come to power. Diesel generators (DG) are used to meet the necessary burden in remote areas. DG's work expenses are high as oil derivative expenses and generator support change. Then, alongside its different half races exchange vitality including photovoltaic and wind.

Techno-financial analysis of PV projects that serve as a decentralized energy source in remote India was conducted by Nouni *et al.* (2006) [3]. In the light of the convection of thermal energy to cooling water [4], the energy conversion efficiency of the PV system was increased by 15% during instant solar radiation. For these cases the entire PV cluster is mounted inside the glazing and consists of an absorber and a heat moving liquid, such that the heat energy absorbed by the PV panel may be used by the hot air or water. The heat collected can be used for hot air and hot water needs at home [5]. Before incorporating them into building structures or lighting the village with decentral electricity, it is necessary to optimize renewable energy sources. In general, batteries are used to resolve discord between output and supply through auto-sufficient renewable sources [6, 7]. Battery-powered batteries have generally been used for off-frame renewable energy studies [8] due to their short time required for the development of an energy bank and an adaptable installation area. The low self-discharge (less than 0.3%) and the low cost and decent cycle performance (63–90%) [9, 10] are consistently the best option for smaller grids for each battery supplying the lead acid and lithium-ion batteries [10, 11] due to their quick and reliable sensitive time. A hybrid energy source was developed by Sorrentino *et al.* [12] with the use of a fuel cell hybrid power optimization tool. The study of technology and feasibility has been published.

Amutha and Rajini [13] evaluated central, modern and farm burden in South India's rural areas and identified an acceptable option for the design of HOMER. They believed that the hybrid PV/wind/hydro/battery configuration was a viable, technologically monetary and green alternative compared to the increased grid. In 12 regions in northern India, Sinha and Chandel [14] have been using the HOMER method to explore the small wind-based hybrid network. Furthermore, they spread the fixed sun and tilting in a low-blusters, Indian uneven territory with a large solar asset, following a PV small-scale wind hybrid power network. Kolhe *et al.* [15] examined how the Sri Lankan rustic electrification hybrid off-grid sustainable power source worked and presented an ideal hybrid cross-section of the solar, diesel and battery systems used. Bhandari *et al.* [16] conducted an experiment on two off-grid hybridizations in the Makawanpur Area of Nepal involving photovoltaic, wind and small hydro hotspots. On the basis of the hourly Adrar (Southern Algeria) calculation, Bagdadi [17] investigated the efficiency of the hybrid PV – Wind – Diesel – battery system. By incorporating it into the exploratory windows Karthick *et al.* [18–20] studied the efficiency of the semi-transparent photovoltaic module (STPV). The efficiency of the room is compared to the STPV module and the DCG window. The coordination of the system for renewable power sources in buildings has been shown to be a feasible arrangement, and in addition to energy generation, it decreases cooling charge. The financial practicalities of hybrid energy systems like solar, wind and diesel generators, which rely on the cost of energy (COE) for use in remote areas of Southern Ghana, and net present costs for the projected frameworks used by Logger have been explored by Adaramola *et al.* [21]. The economy of each system is analyzed with respect to NPC-based zone, diesel utilization and sustainable division. The potential of Punjab wind, photovoltaic and biomass assets was investigated by Jameel Ahmad *et al.* [22]. In Kallar Kahar, sustainable energy sources with imaginable districts and outside projects are identified as being a potential site for the development. The HOMER tool has been used to update the entire 50MW hybrid power station system. In order to learn a better performing system between the constant speed diesel and the variable speed generators, with or without PV, the author presented the same study. The evaluation of hybrid systems comprising photovoltaic, a wind turbine, a battery, an electrolyzer and a hydrogen tank was streamlined by AblaKhairuddin *et al.* [23]. The development of the system depends on the utilization of cost efficiency, energy model for evaluating segment stress and the logical development iterative calculation. The use of fuel cells may be believed to be shocking in a stand-alone system. The efficiency of the integrated photovoltaic module with and without phase change material was investigated by Karthick *et al.* [24]. Eutectic blends have the ability to remove heat and increase the system's efficiency under hot and humid conditions. They were concluded. In eleven Sub-Saharan African countries, a recent survey of around 4,000 clinics and hospitals was conducted by the World Health Organization (WHO) [25]. One-quarter of the health facilities at the site in question were found without power and about three-quarters had an inadequate energy supply. This has frequently been the situation for the supply of epileptic power (which characterizes inadequacy generations), also in national grid-linked installations. Such off-grid clinics and hospitals have historically been used to fuel and install inadequate energy supplies in grid-connected buildings, contributing to an immense price of diesel, unequal distribution, and the high CO emissions that contribute to air pollution and climate change exposure.

A review of literature on rural electrification indicates that combining renewables is one of the best ways to power these rural areas. The combination of renewable energies (RES) approach has shown that electricity is high quality and effective for various rural applications [26, 27]. The aim of the work is to dissect the technical and monetary review of the hybrid sustainable energy source asset for a health care building in the context of the expectation of the hybrid energy source capacity of zero energy buildings in the smart city. In addition to that this analysis will help researchers to design and solve the remote electrification of the rural village with the hybrid energy sources.

8.2 Objectives and Methodology

The main goal is to assess building efficiency by incorporating renewable sources of energy. A structural methodology is built for the study.

The HOMER simulation method is used to construct the device by calculating the electric loads used in health care buildings and health care establishments for everyday activities. The geographical location of the site is shown in Figure 8.1. This is the approach used in the analysis.

- The selection for the Chennai location of optimal hybrid energy sources should be established.
- For computational purposes, the HOMER simulation tool is used. In addition, the HOMER simulation method is better: to build and analyze hybrid energy sources' efficiency over the year.
- For health care facilities for charging the hybrid device, the cost of electric appliances is estimated. Table 8.1 indicates the approximate load.

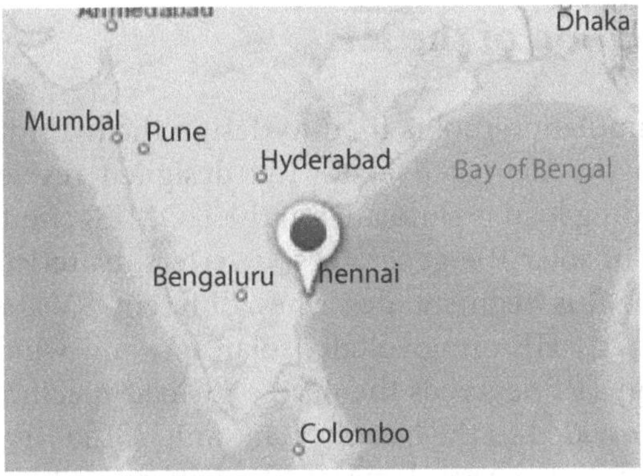

Figure 8.1 Geographical location of the test site.

Table 8.1 DC appliances.

Appliances	Load (W)	Total energy (KWh)
Air conditioner	1050	10.5
Refrigeration	600	7.2
Lighting system	20	0.6
Ceiling fan	30	0.6
Dc water heater	500	0.25
DC pump	750	0.75

- A simulation and financial study will be conducted using the alternative energy source and the battery device.

The hybrid energy sources play a vital role in the zero-energy building. The solar energy is an intermittent. Integration of the renewable energy into the building will take the building from energy consumer to energy producer. Installation of the hybrid energy source is costly. It should be analyzed before installing it into the roof or building envelope. The various simulation software is available in the market. The HOMER is found to be a feasible rate and user friendly. In this work the hybrid energy system is designed to match the energy requirements of the health care zero-energy building. The energy consumption in its day-to-day life is estimated. The energy requirements of the building DC load is shown in Table 8.1.

8.3 Description of the HE

The HOMER simulation tool is used to design the hybrid energy sources with respect to the estimated load. The designed residential load and health care building load is simulated in the HOMER simulation tool. The HE is made up of solar PV, generators, inverters, batteries and a number of components. This segment describes different PVs, wind combinations available, e.g., HE photovoltaic, solar, PV, and wind power. When renewable energy (RE) exceeds the necessary load specifications, an extra power will be provided to the battery bank unit. If the RE generation isn't enough to satisfy the charge demand, the battery will be activated. PV/Wind HE activity is focused on each component's output. PV – Framework for Wind Energy. PV electricity is abundant and emissions-free worldwide. It depends on the location of the power output of the PV system. The PV network is a possible RE outlet for fossil fuel dependency. A hybrid generator of PV, DG, inverter and battery is shown in Figure 8.2 in the window.

The various combinations of hybrid PV/diesel systems via battery bank and additional units DG will be taken into consideration to explore potential benefits and usage of PV-diesel. In the atmosphere, wind is also free of emissions, energy sources. The efficiency of the wind power generation system depends on the wind capacity of the design location. The choice of a higher geographical position for the potential for wind power is a requirement for a viable wind solution and a technologically economic wind power system. In terms of average speed and turbine characteristics, the feasible and economical models for wind turbine systems compare and evaluate the output of different wind turbines at different locations.

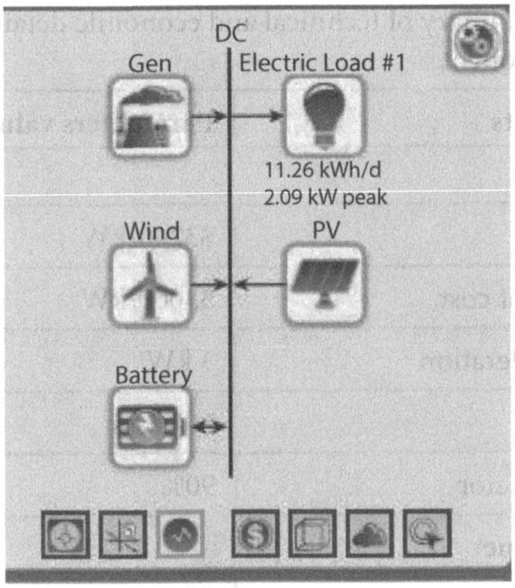

Figure 8.2 Hybrid Energy system in HOMER software.

Renewable PV and wind services are mutually compatible in terms of environmental conditions. As a result, the combined use of PV and wind hybrids is more efficient year-round in energy generation compared to individual sources. Single-source systems, such as photovoltaic or wind energy systems, require the support of stand-alone batteries and DG units to improve the efficiency of hybrid system costs, reliability, energy quality and environmental emissions. In order to optimize the design of PV-based hybrid systems, a number of system performance parameters, such as total annual costs, loss of energy supply probability, responsive analysis, demand side management, loss reductions and polluting emissions, are developed. Table 8.2 shows the summary of the renewable energy source used for the study.

8.4 Results and Discussion

The performance of the solar energy applications depends on the geographical location, incident solar radiation and the environmental parameters. Hence in this study the solar radiation, wind velocity, ambient parameters and the PV, wind energy output are measured. The output power and the energy consumption pattern is discussed in the subsequent section. Figure 8.3 presents the global kWh/m^2/day for Chennai City horizontal radiation. In June and December, the highest and lowest intensity is 6.75 kWh/m^2/day and 2 kWh/m^2/day. When solar radiation is high, DG is used to shore

Table 8.2 Summary of technical and economic details of the system component.

Components	Parameters values
PV array	
Capital cost	\$3200/kW
Replacement cost	\$3000/kW
Sizes consideration	3 kW
Lifetime	20 years
De-rating factor	90%
Wind turbine	
Model BWC WL.1	
Rated capacity 1 kW	
Initial cost per unit	\$5725
Replacement cost	\$3650
Maintenance cost	\$100/year
Units consideration	3 kW
Lifetime	20 years
Battery	
Model Trojan T-5	
Rating	6 V, 225 Ah, 1.35 kWh
Initial cost per unit	\$174
Replacement cost	\$174
Maintenance cost	\$5/year
Lifetime	5 years
Converter	
Capital cost	\$200/kW
Maintenance cost	\$10/year

(Continued)

Table 8.2 Summary of technical and economic details of the system component. (*Continued*)

Components	**Parameters values**
Sizes consideration	0, 1, 2, 3 & 4 kW
Lifetime	10 years
Efficiency	90%
Diesel generator	
Rating	2.6 kW, 20 A
Maximum load ratio	10%
Initial cost per unit	$200/kW
Replacement cost	$200/kW
Operational cost	$0.5/h
Lifetime	15,000 h
Sizes considered	0, and 3kW
System economic	
Interest rate	8%
Inflation rate	2%
Diesel price	$1.1/L

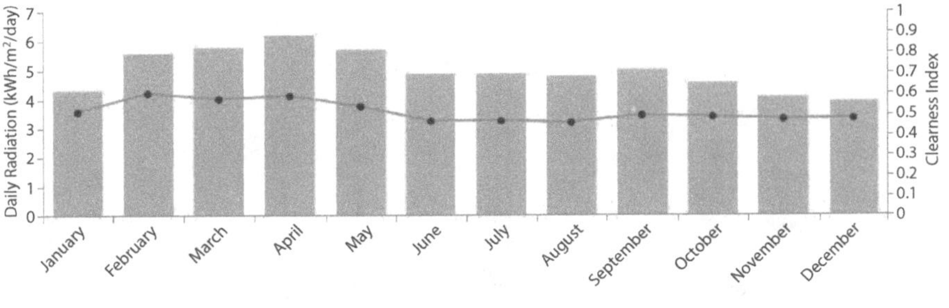

Figure 8.3 Global solar radiation at Chennai 2019.

up the hybrid power supply network. Also, in this sense, before it is constructed on the ground, the position and construction of the ground, such as production, output and capacity, must be measured. At a time when wind speed is not reasonable (low), generators produce only the full power at a slightly faster rate.

Figure 8.4–8.5 show potential wind velocity and average ambient temperature of Chennai city. In this figure, it may be seen that June to August have the highest pace of wind speed month to month. Uncovers the graph of flat radiation relating for Chennai city. As is clear, the most elevated and least degree of solar radiation with 225 kWh/m^2/day and 150 kWh/m^2/day, happened in April and July respectively. Figures 8.6-8.7 shows the month-to-month normal power generation by utilizing the photovoltaic gatherers; 11,823 kWh/yr and wind turbine 637 kWh/yr were acquired for Chennai city.

The generator likewise delivered 3771kWh/yr energy for Chennai city. Figures 8.6 – 8.7 display the power output of the PV and wind systems. The chart shows that electricity is created during the day. During the day, the color changes. During nighttime, there is no power generation, and when

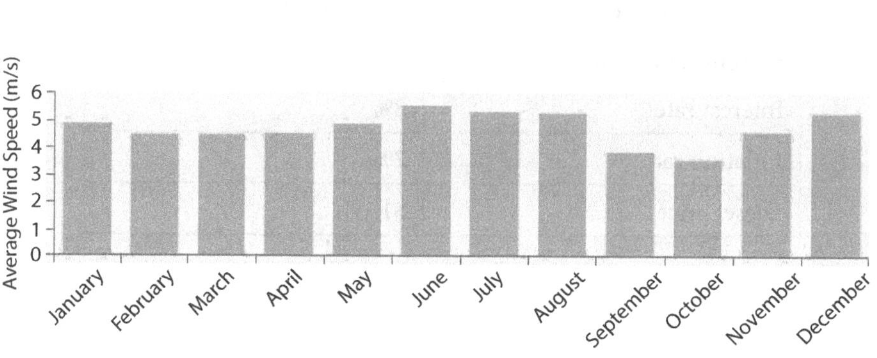

Figure 8.4 Wind resource assessment at Chennai.

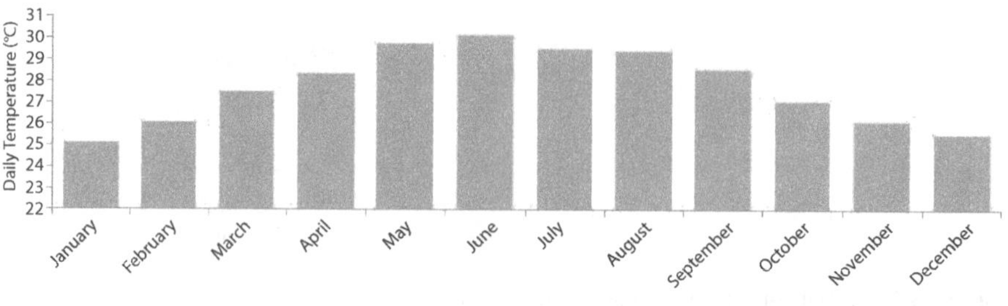

Figure 8.5 Average daily temperature of the site.

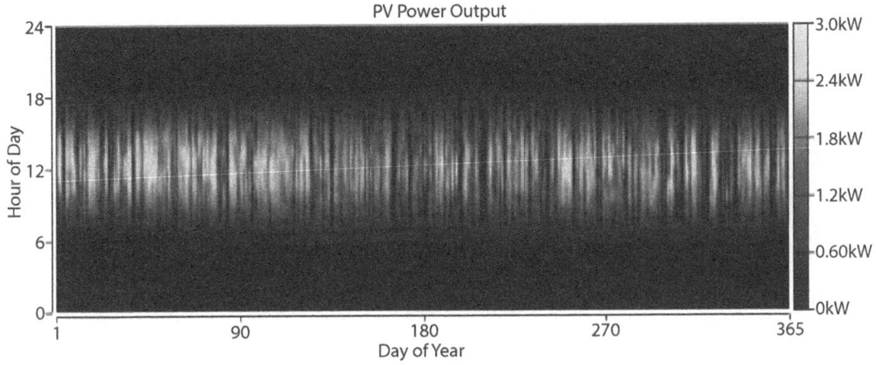

Figure 8.6 Output power of the photovoltaic module.

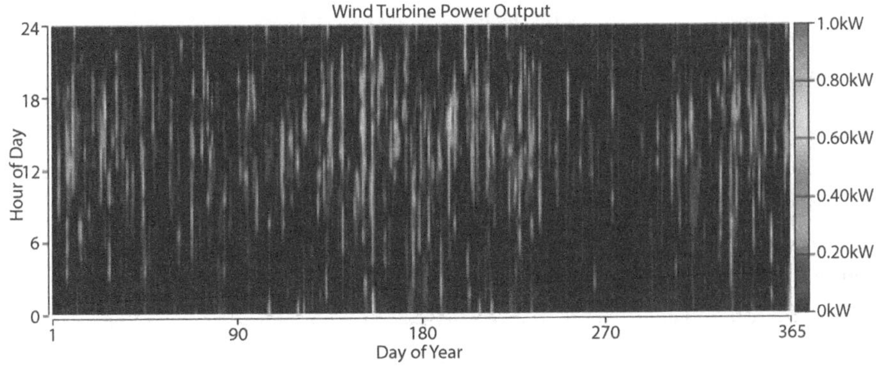

Figure 8.7 Output power from the wind generator.

there is no power generation, it is always dark. The wind turbine's power generation potential is up to 1 KW daily. The energy generated by the wind potential in the selected location is higher during the year.

The average working period for the diesel engine is determined by the availability of solar radiation, summer, winter or windy seasons in Figure 8.8. The graph reveals that during summer, rainy days and during winter days the diesel generator is at least and higher. During the charging or discharge stages, Figure 8.9 shows the battery operation and charging status. The color indicates the difference in loading and unloading capacity. The battery is charged during the day and drained during hours of non-sunshine. Figure 8.10 shows PV modules and generators regularly generated from month to month. In just a few months, the power produced by the generator will have a critical commitment. For this analysis, the city of Chennai was purposely chosen for a total of 3 kW PV Array,

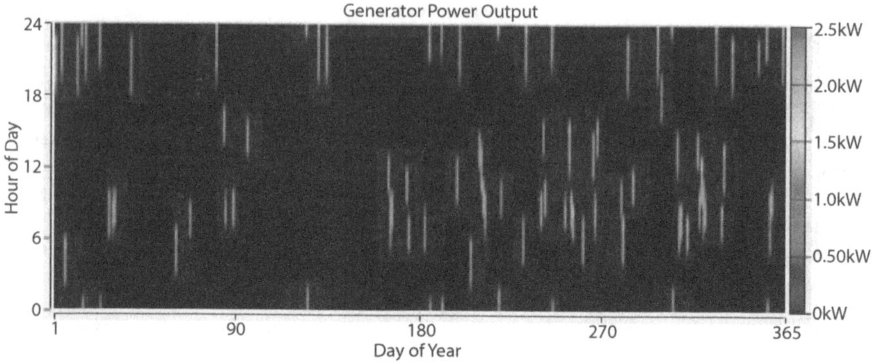

Figure 8.8 Output power from the DC generator.

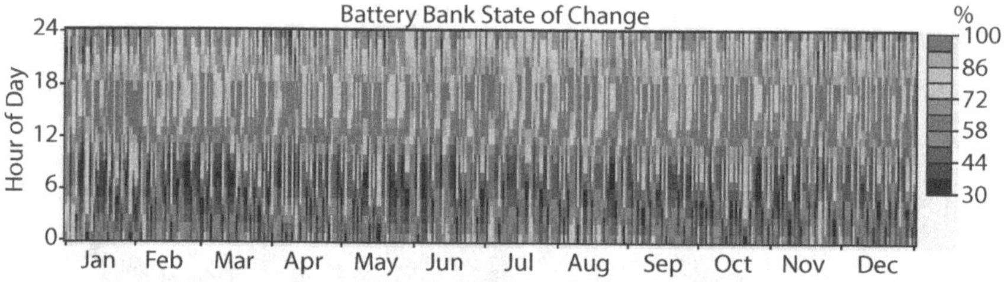

Figure 8.9 Battery operation of the hour of the day.

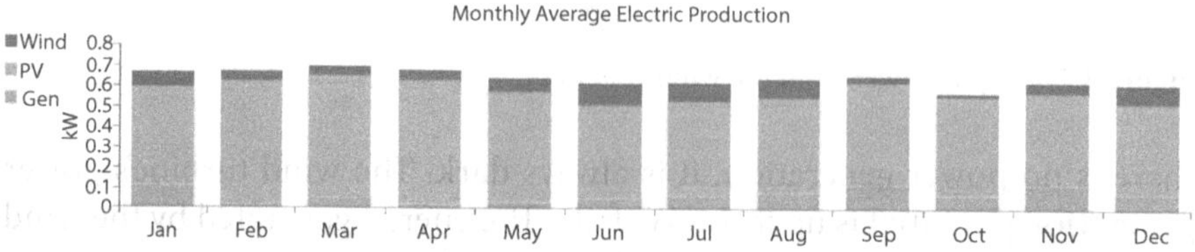

Figure 8.10 Monthly average electricity produced from hybrid energy resources.

3 kW Non-enclosure wind turbines, a 2.8 kW DC Generator and a 3 kWh Hoppecke battery. Gross cash costs of 604,696 dollars. HOMER tool calculates a charged expense of vitality of $5,877/kWh and a work budget of about $63,885/yr. Figure 8.10 displays hybrid frame yield intensities such as PV, wind turbine, DC, battery. The PV output range of the wind turbine is 0.1 to 5.04 hours per day, while the wind turbine is up to 2.4 kw and the DC generator is up to 2 kW.

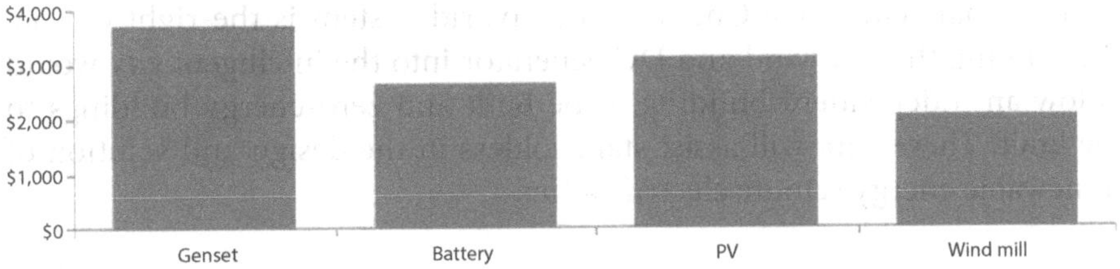

Figure 8.11 Summary of the cost of the component of hybrid systems.

Figure 8.10 displays the complete power generation from the PV-Wind-DG. The efficiency of the device and availability of solar radiation from renewable energy sources are proving to be the basis for efficient energy production. The machine optimization shows that PV and DG operations are the best configuration because of the solar energy availability. This shows that the full power of the photovoltaic system and the diesel generator and the wind turbine produce at least the minimum of energy at times. Although the wind capacity is available all year round, the maximum wind production is small. The wind turbine system is therefore eliminated. The process of all the sources will be streamlined in the second configuration. Figure 8.11, which shows the summary of the cost of the component of the PV-Wind-Diesel-Battery-Converter System, exhibits a thorough investigation.

8.5 Conclusion

The HOMER optimization tool estimates the capacity of the hybrid energy system for the Chennai intelligent city area.

- The proposed hybrid power system provided the necessary power to meet demand during the summer day.
- Month-to-month standard photovoltaic modular power generation is purchased from the City of Chennai at the rate of 11.823kWh/yr and 637kWh/yr.
- The wind turbine worked and provided the energy needed to use the energy available on a windy day.
- This generator also has a capacity for Chennai City of 3771 kWh/yr. The peak and lowest solar radiation rates were evident in April and July, with 225 kWh/m^2/day and 150 kWh/m^2/day respectively.

For smart cities like Chennai, the Hybrid System is the right option. Integrating the PV, wind and DC generator into the intelligent city would allow an independent building to be built and zero-energy buildings to be built. These data will assist stakeholders in the design and solution of renewable energy remote electrification.

Nomenclatures

PV	Photovoltaic
DG	Diesel generator
HE	Hybrid energy
HOMER	Hybrid Optimization Model for Electric Renewables
DC	Direct Current
STPV	semi-transparent photovoltaic module
WHO	World Health Organization
COE	Cost of Energy

References

1. Reddy K, P., Gupta, M. V. N., Nundy, S., Karthick, A., & Ghosh, A. (2020). Status of BIPV and BAPV System for Less Energy-Hungry Building in India—A Review. *Applied Sciences, 10*(7), 2337.
2. Karthick, A., Kalidasa Murugavel, K., Kalaivani, L., & Saravana Babu, U. (2018). Performance study of building integrated photovoltaic modules. *Advances in Building Energy Research, 12*(2), 178-194.
3. Nouni, M. R., Mullick, S. C., & Kandpal, T. C. (2006). Photovoltaic projects for decentralized power supply in India: A financial evaluation. *Energy Policy, 34*(18), 3727-3738.
4. Karthick, A., Murugavel, K. K., & Prabhakaran, D. S. R. (2017, March). Energy analysis of building integrated photovoltaic modules. In *2017 International conference on power and embedded drive control (ICPEDC)* (pp. 307-311). IEEE.
5. Chow, T. T., Ji, J., & He, W. (2007). Photovoltaic-thermal collector system for domestic application.
6. Krishnavel, V., Karthick, A., & Murugavel, K. K. (2014). Experimental analysis of concrete absorber solar water heating systems. *Energy and Buildings, 84*, 501-505.
7. Ramanan, P., & Karthick, A. (2019). Performance analysis and energy metrics of grid-connected photovoltaic systems. *Energy for Sustainable Development, 52*, 104-115.

8. Karthick, A., Kalidasa Murugavel, K., Sudalaiyandi, K., & Muthu Manokar, A. (2019). Building integrated photovoltaic modules and the integration of phase change materials for equatorial applications. *Building Services Engineering Research and Technology*, 0143624419883363.
9. Javed, M. S., Song, A., & Ma, T. (2019). Techno-economic assessment of a stand-alone hybrid solar-wind-battery system for a remote island using genetic algorithm. *Energy, 176*, 704-717.
10. Ramanan, P., Kalidasa Murugavel, K., Karthick, A., & Sudhakar, K. (2020). Performance evaluation of building-integrated photovoltaic systems for residential buildings in southern India. *Building Services Engineering Research and Technology, 41*(4), 492–506. https://doi.org/10.1177/0143624419881740
11. Khare, V., Nema, S., & Baredar, P. (2016). Solar–wind hybrid renewable energy system: A review. *Renewable and Sustainable Energy Reviews, 58*, 23-33.
12. Sorrentino, M., Adamo, A., & Nappi, G. (2019). Optimal sizing of an rSOC-based renewable microgrid. *Energy Procedia, 159*, 237-242.
13. Amutha, W. M., & Rajini, V. (2016). Cost benefit and technical analysis of rural electrification alternatives in southern India using HOMER. *Renewable and Sustainable Energy Reviews, 62*, 236-246.
14. Sinha, S., & Chandel, S. S. (2015). Prospects of solar photovoltaic–micro-wind based hybrid power systems in western Himalayan state of Himachal Pradesh in India. *Energy Conversion and Management, 105*, 1340-1351.
15. Kolhe, M. L., Ranaweera, K. I. U., & Gunawardana, A. S. (2015). Techno-economic sizing of off-grid hybrid renewable energy system for rural electrification in Sri Lanka. *Sustainable Energy Technologies and Assessments, 11*, 53-64.
16. Bhandari, B., Lee, K. T., Lee, C. S., Song, C. K., Maskey, R. K., & Ahn, S. H. (2014). A novel off-grid hybrid power system comprised of solar photovoltaic, wind, and hydro energy sources. *Applied Energy, 133*, 236-242.
17. Baghdadi, F., Mohammedi, K., Diaf, S., & Behar, O. (2015). Feasibility study and energy conversion analysis of stand-alone hybrid renewable energy system. *Energy Conversion and Management, 105*, 471-479.
18. Karthick A, Ramanan P, Ghosh A, Stalin B, Vignesh Kumar R, Baranilingesan I. Performance enhancement of copper indium diselenide photovoltaic module using inorganic phase change material. *Asia-Pac J Chem Eng*. e2480 2020. https://doi.org/10.1002/apj.2480
19. Karthick, A., Murugavel, K. K., & Ramanan, P. (2018). Performance enhancement of a building-integrated photovoltaic module using phase change material. *Energy, 142*, 803-812.
20. Karthick, A., Murugavel, K. K., & Kalaivani, L. (2018). Performance analysis of semitransparent photovoltaic module for skylights. *Energy, 162*, 798-812.
21. Adaramola, M. S., Agelin-Chaab, M., & Paul, S. S. (2014). Analysis of hybrid energy systems for application in southern Ghana. *Energy Conversion and Management, 88*, 284-295.

22. Akhmad, K., Kitamura, A., Yamamoto, F., Okamoto, H., Takakura, H., & Hamakawa, Y. (1997). Outdoor performance of amorphous silicon and poly-crystalline silicon PV modules. *Solar Energy Materials and Solar Cells, 46*(3), 209-218.
23. Khiareddine, A., Salah, C. B., Rekioua, D., & Mimouni, M. F. (2018). Sizing methodology for hybrid photovoltaic/wind/hydrogen/battery integrated to energy management strategy for pumping system. *Energy, 153*, 743-762.
24. Karthick, A., Murugavel, K. K., Ghosh, A., Sudhakar, K., & Ramanan, P. (2020). Investigation of a binary eutectic mixture of phase change material for building integrated photovoltaic (BIPV) system. *Solar Energy Materials and Solar Cells, 207*, 110360.
25. Adair-Rohani, H., Zukor, K., Bonjour, S., Wilburn, S., Kuesel, A. C., Hebert, R., & Fletcher, E. R. (2013). Limited electricity access in health facilities of sub-Saharan Africa: a systematic review of data on electricity access, sources, and reliability. *Global Health: Science and Practice, 1*(2), 249-261.
26. Roy, C., Motamedi, S., Hashim, R., Shamshirband, S., & Petković, D. (2016). A comparative study for estimation of wave height using traditional and hybrid soft-computing methods. *Environmental Earth Sciences, 75*(7), 590.
27. Karthick, A., Manokar Athikesavan, M., Pasupathi, M. K., Manoj Kumar, N., Chopra, S. S., & Ghosh, A. (2020). Investigation of Inorganic Phase Change Material for a Semi-Transparent Photovoltaic (STPV) Module. *Energies, 13*(14), 3582.

9

Hybrid Solar-Biomass Gasifier System for Electricity and Cold Storage Applications for Rural Areas of India

Nasir ul Rasheed Rather[1]* and Umakanta Sahoo[2]

[1]*CSIR-Central Mechanical Engineering Research Institute, Durgapur, West Bengal, India*
[2]*National Institute of Solar Energy, GOI, Gurugram, Haryana, India*

Abstract

This chapter mainly deals with performance evaluation of hybrid cold storage using solar & exhaust heat of biomass gasifier. The system consists of a 50 kWe biomass gasifier including gas cleaning and cooling system, 50 kWe 100% producer gas engine, heat recovery unit, 15 kW (NH_3-H_2O) vapor absorption cooling system, 4 Scheffler dishes with 16 m^2 area for each and cold storage room of capacity of 20 metric tons. The 50 kW biomass gasifier system provides electricity and the waste heat of the engine (exhaust) that is used as the main source of energy for the cold storage. During the daytime, when sunshine is available, the Scheffler dishes along with engine exhaust provide heat to the system. During evening, when there is no sunlight, the gasifier is run to meet the evening electricity load and the cold storage will operate only on engine exhaust. In case no sunshine is available and engine exhaust is also not sufficient, provision has been made for firing of producer gas in a heat recovery system (HRS) to meet balance heat requirement for running the cold storage. Study is carried out for automation of process in different load condition. The study has been undertaken to optimize the system performance during different radiation and ambient conditions. The maximum COP of the system was found 0.60 at Generator temperature of 85.91°C. The COP of the system (VAM) increased with the increase in Condenser Temperature. The highest condenser temperature obtained was 26.83°C at COP of 0.60. The minimum evaporator temperature of - 4.24°C was achieved.

**Corresponding author*: nasirrasheed9999@gmail.com

Umakanta Sahoo (ed.) Hybrid Renewable Energy Systems, (199–246) © 2021 Scrivener Publishing LLC

Keywords: Solar, biomass gasifier, cold storage, heat recovery system

9.1 Introduction

India is a predominantly rural country. Notwithstanding present growth rates, it would be a long time before the population in the villages exceeds the population in the urban areas. It is also accepted that urbanization adds much more pressure and expenses in providing civic amenities, especially water and sewage facilities. Enough has been said about the energy demand projection outgrowing the energy generation with respect to electricity. As India's energy demand increases, its contribution to global warming is also likely to increase; however, there is a lot of pressure about India's need to start looking at reduction of greenhouse gases. Thus, we have three independent problems, namely,

- Low Rural economy
- Electrical energy shortage
- Post-harvest losses in agriculture and allied sectors

Thus the major part in India is encountering problems of electricity, irrigation and cold storage infrastructural facilities. The immediate perspective is to make available at least bare minimum energy services to the rural people. The technology, which does the trick convincingly in the remote rural environment, is biomass gasification. Therefore, biomass-based gasifiers power plant and utilization of waste heat (exhaust) of biomass producer gas engine for cold storage may be the most appropriate technology for the development of rural India.

Electricity is the most critical of inputs for establishment of a cold storage and its operation. The chief reason for the absence of cold storages near farms is the unavailability of electricity in villages, besides the manner in which business is conducted around cold storages. Although India boasts of 95% electrification in villages, only around 55% of households in rural India are electrified, and the ones which are fortunate enough to be electrified receive electricity for only few hours a day (around six hours a day during off-peak period), and that too with critical uncertainty. In a conventional cold storage, the energy expenses account for approximately 28% of costs, which also shows the importance of it in daily operation of the cold storages. The reason that not many private or government enterprises have entered the cold storage market or installed cold storages near farm gates or in villages is because they do not see much business sense in it. This is mainly due to the absence of grid electricity. The only alternate way therefore to run the cold storages is by means of diesel engines, which can prove to be extremely uneconomical, especially when preserving a lower quantum of produce, (covering a village or few at most). In this particular innovative solution, the cold storage is run chiefly on the thermal cycle—typically used in outsized cold storages or air conditioning systems—instead of electricity. Together with the system design, a scoping study was carried out to understand the status of cold storage requirement across pan-India. In this study various aspects were included: statewide production quantum of fruits and vegetables, supply chain and the role of various actors in it, wastages, availability of cold storage facilities, storage and operational method used in conventional large-scale cold storage, status and demand of cold storages in rural areas, etc. This has greatly helped in forming the base of this project and rationalizing its requirement. These interventions can lead to significant development in rural areas by providing both electricity and cold storage facility which has visible impacts on overall socioeconomic development in rural areas. Other benefits include introducing entrepreneurial wisdom, providing an alternative to burning kerosene for lighting, and most importantly in improving the quality of life of the villagers.

Keeping in view the above facts, a Solar-Biomass hybrid model for power generation and cold storage for rural areas will be developed having the following objectives:

- Performance evaluation of a Solar-Biomass model for Power generation and Cold storage.
- To determine the Coefficient of performance (COP) of vapor Absorption Machine (VAM) used in cold storage and to see the effect of COP on generator temperature, evaporator temperature and condenser temperature of VAM.
- To study the relationship between auxiliary firing and mass of wood.
- Energy conservation by using the waste heat (engine exhaust) for cold storage.
- Economic analysis of the system as compared to conventional electricity based cold storage system.

9.2 Literature Review

9.2.1 Gasification of Biomass

Rajiv Varshney *et al.* (2010) observed a lot of surplus biomass is available in different parts of the country. The pine tree needles which cause forest fires in the Himachal Pradesh can be utilized directly as a fuel, briquetting or for gasification. This can lead to saving of immense forest wealth from fires. Haryana being an agricultural state has a huge potential of biomass availability in the form of crop residue and sawdust. In the agricultural sector, a total 24.697 Mt y-1 of the residue is generated, of which 71% is consumed in various domestic and commercial activities within the state. In the agro-based industrial sector a total of 646 kty-1 of sawdust is generated out of which only 6.65% is consumed in the state. Of the total generated biomass in the state, 45.51% is calculated as basic surplus, 37. 48% as productive surplus and 34.10% as net surplus. The power generation potential from all these three categories of surplus biomass is 1.499 GW, 1.227 GW and 1.120 GW, respectively.

Jiu Huang Klaus *et al.* (2011) highlight the Syngas prevention. The high residual tar concentrations in syngas prevent any ambitious utilization. At a constant flow rate of 20 Nm3/h and temperatures of 500°C, 600°C and 700°C the tar conversion rates reached 44.9%, 78.1% and 92.3%, respectively. These results could be increased up to 98.6% and 99.3% by using an operating temperature of 700°C and lower flow rates of 15 Nm3/h and 10 Nm3/h. The syngas quality after the purification process at 700 °C/10 Nm3/h is acceptable for inner combustion (IC) gas engine utilization.

Worleyand *et al.* (2012) reported on Gasifier Technology Assessment. In Golden, Colorado, the National Renewable Energy Laboratory (NREL) recently published several studies on thermochemical conversion of biomass for the production of ethanol via gasification. These studies include:

- "Thermochemical Ethanol via Indirect Gasification and Mixed Alcohol Synthesis of Lignocellulose Biomass," the production of ethanol via indirect gasification of biomass based on a Battelle Columbus Laboratory (BCL) gasifier design.
- "Thermochemical Ethanol via Direct Gasification and Mixed Alcohol Synthesis of Lignocellulose Biomass," describing the production of ethanol via direct gasification of biomass using an Institute of Gas Technology (IGT) gasifier design.
- "Techno-economics of the Production of Mixed Alcohols from Lignocellulose Biomass via High Temperature Gasification,"

describing the production of ethanol via entrained flow slagging gasification of biomass. These reports demonstrate that there are great opportunities to apply various gasifier technologies in the conversion of biomass to syngas for the production of renewable fuels. Each of these reports shows that a substantial portion of a project's capital cost is attributable to the gasifier and that the overall project cost increases with gasifier design and operational complexity. The cost values used in these reports for the gasifier economics are based on the small amount of data available in the literature but improve confidence in the predicted economics of these technologies, although a thorough understanding of the basic capital cost and engineering requirements for gasifiers is necessary.

Younes Chhiti (2013) noted that the possibility of using this gas for heating and power generation was soon realized and then emerged in European producer gas systems, which used charcoal and peat as feed material. At the turn of the 20th century petroleum gained wider use as a fuel, but during both world wars and particularly World War II, shortage in petroleum supplies led to widespread reintroduction of gasification. By 1945 the gas was being used to power trucks, buses and agricultural and industrial machines. It is estimated that there were close to 9,000,000 vehicles running on producer gas all over the world.

9.2.2 Solar Energy Cooling and Heating

Papadopoulo *et al.* (2010) did an analysis of solar cooling, focusing on the state of the art of thermal solar systems use and on the possibilities of combining those with state-of-the-art technologies in absorption refrigeration, in order to cover the cooling demand of residential and commercial buildings. This was done by assessing the available solar and refrigeration technologies as well as by highlighting the situation in the building market, as this was still the dominant factor for the propagation of such systems.

Rodriguez *et al.* (2010) performed a study on flat plate thermal solar collector efficiency and transient behavior under working condition. Three thermocouples were added to measure the center point temperature of the glass cover, box back surface and absorber plate of one of the collectors. Using this information, the parameters of a detailed thermal network model were determined. By using this model useful heat and thermal losses were calculated by applying a dynamic energy balance under the transient regime. The facility was based on a nine-year-old on-campus field with 50 m^2 flat plate solar collectors, which operated for a domestic hot water (DHW) application. The working parameters were recorded during an entire year for periods of 10 min. The model was experimentally validated by comparing its results to the instant collector temperatures and heat fluxes that were obtained from the experimental database. Both of the experimental and model results were compared to the direct application of the collector efficiency normalization curve (ENC). Since the operating conditions were unlike, differences throughout the day became evident. The influence of several factors was fully described.

Wang *et al.* (2012) investigated the temperature field and thermal stress of dish solar concentrator's focal region. With finite element method, they used a modified radiant heat flux function dependent on the solar radiant flux and velocity, to calculate the temperature field of a plate on the focal region of a dish solar concentrator. There was a good accordance when these results were compared with pictures taken by an infrared camera.

9.2.3 Engine Exhaust and Waste Heat Recovery

Mago *et al.* (2010) observed environmental emissions are the primary driving forces in the research for the sustainable and economically viable technologies. Improvement in fuel conversion efficiency requires an analysis of various losses encountered in the engine. Analysis for the waste heat recovery were based on the use of organic rankine cycles. This paper demonstrated the performance analysis of the different components of the rankine cycle and minimum losses measured of components gave higher performance.

Conklin (2010) stated that in internal combustion engines a huge amount of energy is lost in the form of heat through the exhaust gas. The study found that the percentage of fuel energy converted to useful work was only 10.4% and also found the thermal energy lost through exhaust gas was about 27.7%. The second law (i.e. Energy) analysis of fuel has been shown that fuel energy is converted to the brake power about 9.7% and the exhaust about 8.4%.

P. Sathiamurthi (2011): This heat depends in part on the temperature of the waste heat gases and mass flow rate of exhaust gas. Waste heat losses arise both from equipment inefficiencies and from thermodynamic limitations on equipment and processes. For example, consider the internal combustion engine, in which approximately 30 to 40% is converted into useful mechanical work. The remaining heat is expelled to the environment through exhaust gases and engine cooling systems.

Kumar *et al.* (2011) concluded that a major part of the heat supplied in the IC Engine is dumped into the atmosphere as waste heat. If this waste heat energy is tapped and converted into useable energy, the overall efficiency of an engine can be improved. The percentage of energy rejected to the environment is approximately 30-40% of the energy supplied by the fuel depending on engine load. Thermoelectric modules which are used as thermoelectric generators are solid-state devices that are used to convert thermal energy from a temperature gradient to electrical energy; it works on the basic principle of Seebeck effect. This paper demonstrates the potential of thermoelectric generation and performance of thermoelectric generators under various engine operating conditions. The study also revealed that thermoelectric generators can reduce the size of the alternator or eliminate them in automobiles.

Zhang and Chau (2011) did an analysis of the radiator system, which is used to pump the coolant through the chambers in the heat engine block to avoid overheating. Conversely, the exhaust gas system of an IC engine is used to discharge the expanded exhaust gas through the exhaust manifold. The study reported that presently TEG is mostly installed in the exhaust gas system (exhaust manifold) due to its simplicity and low influence on the operation of the engine.

Wang *et al.* (2012) found that the value of exhaust gases mentioned to be 18.6% of total combustion energy. It also found that by installing heat exchanger to recover exhaust energy of the engine could could yield up to 34% in fuel saving.

9.3 Materials and Methods

This type of cold storage-cum-power generation hybrid system has been installed at the R&D campus of NISE (National Institute of Solar Energy), Ministry of New and Renewable Energy located at Gwalpahari, Gurgaon. The main aim of this project is to evaluate a solar-biomass hybrid model which will be used for power generation (electricity) and cold storage. This project has also been designed to keep rural development into consideration and will help in sustainable development of rural areas.

9.3.1 System Components

The Solar-biomass hybrid model for power generation and cold storage consists of six major components:

1. Biomass gasifier (capacity: 50 kWe)
2. Gas-engine Generator (capacity: 50 kWe)

3. Waste heat recovery Unit
4. Solar Scheffler dic (4 Scheffler dishes of 16 sq.m. area each)
5. VAM (15 kW)
6. Cold storage unit (capacity: 20 T of vegetable storage)

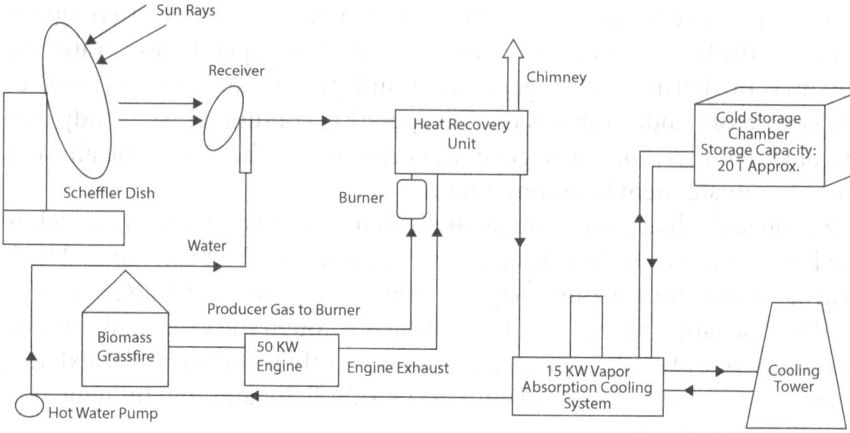

Figure 9.1 Solar Biomass hybrid model for power generation and cold storage model.

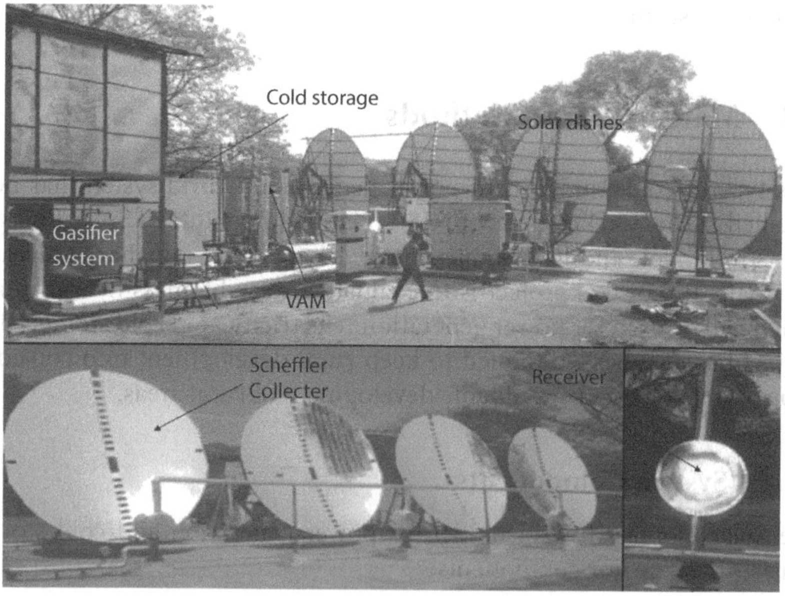

Figure 9.2 Actual setup of Solar Biomass hybrid model for power generation and cold storage.

9.3.1.1 Biomass Gasifier

The complete gasification comprises all the processes which convert the solid fuel into a gaseous and liquid product leaving only parts of the mineral constituents of the fuel as a residue. Biomass is a natural substance available, which stores solar energy by the process of photosynthesis in the presence of sunlight. It chiefly contains cellulose, hemicelluloses and lignin, with an average composition of $C_6H_{10}O_5$, with slight variations depending on the nature of the biomass. Theoretically, the ratio of air-to-fuel required for the complete combustion of the biomass, defined as stoichiometric combustion is 6:1 to 6.5:1, with the end products being CO_2 and H_2O. In gasification the combustion is carried at sub-stoichiometric conditions with air-to-fuel ratio being 1.5:1 to 1.8:1. The gas so obtained is called producer gas, which is combustible. This process is made possible in a device called a gasifier, in a limited supply of air.

Gasification is a two-stage reaction consisting of oxidation and reduction processes. These processes occur under sub-stoichiometric conditions of air with biomass. The first part of sub-stoichiometric oxidation leads to the loss of volatiles from biomass and is exothermic; it results in peak temperatures of 1400 to 1500 K and generation of gaseous products like carbon monoxide, hydrogen in some proportions and carbon dioxide and water vapor, which in turn are reduced in part to carbon monoxide and hydrogen by the hot bed of charcoal generated during the process of gasification. Reduction reaction is an endothermic reaction to generate combustible products like CO, H_2 and CH_4 as indicated below.

$$C + O_2 \longrightarrow CO_2$$

$$C + H_20 \longrightarrow CO + H_2$$

$$C + 2H_2 \longrightarrow CH_4$$

Gasifier components
Gasifier components include:

1. Scrubber
2. Condenser
3. Cyclone
4. Char filter
5. Filters
6. Blower

7. Fabric Filter
8. Pannel Board

Presently a 50 kW wood-based downdraft gasifier is being used for the project. The downdraft or co-current gasifier is prominently used since the tar entrainment problem in the gas stream is minimized in co-current or downdraft gasifiers, in which primary gasification air is introduced at or above the gasifier oxidation zone and the producer gas leaves from the bottom of the gasifier (so, the air and fuel flow move in the same direction – co-current). Tarry distillation products from the fuel in this gasifier types have to pass through the burning bed of charcoal and by means of this they are converted into gas. The tar decomposition degree depends on the gasifier hot zone temperature and on the residence time of tarry vapors there. Wood-based chips are used and fed from the top of gasifier. The moisture content of feed is about 18%. The different type of wood chips are used in the gasifier, depending on availability. The biomass fuel is fed at the top of the gasifier and air is supplied at regular intervals. The hot combustion gases are passed through the fuel bed and this leads to generation of combustible gas.

Advantages of Biomass Gasification

Provides a sustainable and affordable alternative to fossil fuel–based power plants at low power levels. This technology can assure "continuous power supply" even at Peak load conditions.

- Efficiency of the system is very high when compared to other Renewable energy systems such as wind and Solar.
- Entire project is eco-friendly as it is "CO_2 neutral" and generates very little SO_2 & Nitrogen Oxides compared to conventional fossil fuel–based power plants. We can have the benefit of "Green Power".
- Improves country's energy self-reliance and reduces crippling oil import bill.
- It is a cost-effective solution as it combines low Unit Capital Cost with low Unit Cost of Production. Cost of installation per kW is about the same as for large power levels.
- Plant Load factor is comparable to large thermal power plants.

- The process is very energy efficient as the waste heat in the flue gases can be utilized effectively.
- This process produces Very Low Emission of unburnt primary fuel and no Fly Ash, because solid fuel is subjected to pyrolysis and the gas is passed through cleaning and cooling process to remove particulates.
- Biomass production and conversion activities stimulate local economics by providing employment opportunities to local people and also product and market opportunities for agribusiness.

Selection criteria

The selection of Updraft or Downdraft Gasifier would mainly depend upon the available fuel and the requirements of the application. A downdraft wood gasifier requires wood chips, biomass briquettes or wood-like agro-residues (such as stems, stalks). If fuel flexibility is required and need for fuel switching (say from biomass to coal or from one biomass to another) is anticipated, the updraft gasifier would be suitable.

For gasification of coal, an updraft gasifier is suitable. A downdraft gasifier cannot accept coal. For direct heating applications, such as firing in kiln, furnaces, boilers, etc., an updraft gasifier would make a better choice. However, if the contaminants in fuel gas are likely to affect the end-product quality, the choice of downdraft gasifier makes better sense. There are situations where the effect on end-product quality is critical, but fuel flexibility is also an important consideration. In such situations, an updraft gasifier with gas scrubbing arrangement maybe selected.

9.3.1.2 Gas-Engine Generator

Both type of IC engines, i.e., spark ignition engines and compression ignition engines, can be used with biomass-derived fuels, although the technologies are somewhat different. The spark ignition engines operating on gasoline and kerosene are often preferred for applications where low weight is an advantage and where the shorter lifetime (typically less than 5,000 hours) and lower efficiency (typically about 25% at full power) can be accepted from an economic point of view. This situation often applies when the annual number of operating hours is low, say less than 500.

Compression ignition engines operating on diesel fuel are heavier, and are generally designed for a lifetime of more than 20,000 hours, and show a higher efficiency (typically about 35% at full power). They are preferred for applications where weight is less important and where the gains on lifetime and fuel economy compensate for the higher price of engine.

Another possibility is the conversion of the biomass to a combustible gas (producer gas), which can be used as fuel in the engine. Spark ignition engines can use producer gas without major modification. Compression ignition engines will not operate on this gas only, but can be run on a mixture of gas and suitable liquid fuel. The engine used in the present project runs on combustible gas (producer gas) produced from the wood biomass, when it is burned with less air than is needed to achieve complete combustion. The producer gas consists of combustible components carbon monoxide, hydrogen and some methane mixed with carbon dioxide, water vapor and nitrogen. Before its introduction into the engine the gas must be cleaned to remove dust and tars. Various types of cleaners have been used for this purpose, e.g., cyclones, wet scrubbers, large volumes filled with sawdust, wood wool, pieces of cork or coir fiber. However, in this system it's necessary to clean the inlet manifold of the engine more often. The gas outlet of gasifier is connected with the various downstream systems viz cyclones, flare with valve, engine gas control valve, cyclones and engine shut-off valve. Gas produced in the gasifier is scrubbed and cooled. Cool, clean gas and air is then sucked into the engine through a mixer butterfly consisting of piping and valves arrangement. The gas is cooled in order to avoid condensation in the inlet manifold when the gas (which contains some water vapor) is mixed with ambient air. Cooling of gas will also lead to an improved engine power output, as it will increase the volumetric energy content of the gas-air mixture supplied to engine. The gasifier is started with a Battery (12V) and AC Scrubber Pump. The producer gas then starts the engine on gas mode. A governor-linked control butterfly is provided to vary the gas quantity as per electrical load on the generator, keeping frequency within limits. The shaft of the gas engine is coupled with the generator which produces the electricity. Some portion of the electricity is used in the pump of the cold storage system and rest of the electricity is given to the Smart Grid. From the smart grid, the electricity can be used for other end-use application.

9.3.1.3 *Waste Heat Recovery Unit*

Shell in tube heat recovery system is used for recovery of heat from exhaust of engine. The system has also a provision to fire producer gas as an auxiliary arrangement in case sufficient exhaust and solar heat is not available to meet the requirement of vapor absorption machine. In the system pressurized water is used for the heat observer/heat carrier. The system also receives heat from the solar system through the flow of hot

Figure 9.3 Gas-engine generator.

water. The temperature of engine exhaust was found to be near 400°C after the heat recovery flue gas was going and temperature flue gas/exhaust was observed at the outlet of heat recovery system. As such, 55% of the heat of the exhaust was recovered by the heat recovery system. The burner temperature of auxilarry firing system was nearly 600°C whereas the outlet temperature was 200°C. As such, 66% heat was recovered by the Heat recovery system in case of auxillary firing. The provision of three inlets (engine exhaust, auxiliary firing and solar thermal fluid) makes it one of the efficient systems and helps in energy conservation and waste heat utilization. During the non-solar hours, engine exhaust and auxiliary firing of producer gas are used as hot and cold fluids in the heat recovery unit. Auxiliary firing quantity was inversely proportional to exhaust output/ electrical load consumption.

Waste heat is heat, which is generated in a process by way of fuel combustion or Chemical reaction, and then "dumped" into the environment even though it could still be reused for some useful and economic purpose. The essential quality of heat is not the "amount" but rather its "value". The strategy of how to recover this heat depends in part on the temperature of the waste heat gases. As seen in Figure 9.4 the hot gases are carried through a number of parallel small diameter tubes, while the incoming air to be heated enters a shell surrounding the tubes and passes over the hot tubes one or more times in a direction normal to their axes. Although baffling increases the cost of the exchanger and the pressure drop in the combustion air path, it increases the effectiveness of heat exchange. Shell and tube-type Recuperators are generally more compact and have a higher effectiveness than radiation Recuperators because of the larger heat transfer area made possible through the use of multiple tubes and multiple passes of the gases.

The temp of engine exhaust near 400°C, Exhaust gas has recovered by heat recovey, shell type boiler, in such a way that 55% of the heat was recovered by the HRU. The shell in the tube heat recovery system is used for recovery of heat from exhaust of engine (waste energy). The system has also a provision to fire producer gas as an auxiliary arrangement in case sufficient exhaust and solar heat is not available to meet the requirement of vapor absorption machine (75-140°C). In the system, pressurized water is used for the heat observer/heat carrier. The system also receives the heat from the solar system through the flow of hot water.

Figure 9.4 Waste heat recovery unit.

The heat recovery unit is connected with the internal combustion engine on one side and with the Scheffler Dish on the other side Thus heat from the recovery unit is passed to 15 kw Vapor Absorption Machine which is connected with the cooling tower. So the Vapor Absorption Machine is used to run cold storage chamber whose storage capacity is 20 Ton.

9.3.1.4 Scheffler Dish Collector

A Scheffler dish system consists of a solar concentrator and receiver. The solar rays reflected from the concentrators are focused on the receiver. The concentrator essentially is a reflective mirror of polished glass, metal or metallized film that concentrates light and heat from the sun on a small receiver area, making the energy more concentrated and increasing its heating power.

A Scheffler dish is a small lateral section of a paraboloid, which concentrates the sun's radiation over a fixed focus, with an automatic single axis tracking. A Scheffler dish is made of a number of flat-shaped mirrors or reflective films which are mounted on a structural frame. The dish rotates about north-south axis parallel to Earth's axis to track the sun's movement. The axis of rotation passes through the centre of gravity of the reflector and that's how the reflector always maintains its gravitational equilibrium. The Scheffler reflector also performs a change in inclination angle while staying directed to the sun, in order to obtain a sharp focal point. Focus is fixed at distance of focal length of the paraboloid along the axis of the paraboloid. The receiver at a fixed location captures the concentrated heat and transfers it to water/thermic fluid to generate hot water/hot thermic fluid or high-pressure steam.

A Scheffler dish is used for the low-medium process heat applications. This dish can attain temperature up to 150-200°C as per the specific requirements in industries, commercial and residential complexes, religious places, etc. A typical 16 m2 Scheffler dish has thermal capacity equivalent to 30,000 Kcal/ day to 35,000 Kcal/day depending on the manufacturing precision and DNI on a clear sunny day.

Some of the typical application areas of the Scheffler dish are:

- Food processing
- Community cooking
- Solar-assisted comfort cooling
- Cold storage/refrigeration through VAM
- Pharmaceutical
- Drying and dying

Table 9.1 The important characteristics of these designs are as follows.

Parameter	**16 m²**	**32 m²**
Average Aperture Area	11.65 m²	23.3 m²
Shadow Free Area	35 m²	60 m²
Thermic medium	Steam, hot water, thermic fluid	Steam, hot water, thermic fluid
Delivery capacity	Up to 12 bar pressure, Up to 180 °C temperature	Up to 12 bar pressure, Up to 200 °C temperature
Weight	Minimum 400 Kg	Minimum 800 Kg

The design configuration of an industrial process heat solar system depends on the specific application. It may consist of one or more Scheffler dishes in a system.

Collector

The collector of a Scheffler dish is an assembly of flat-shaped solar grade glass mirrors or Aluminium mirror reflectors arranged on a structural steel framework. The mirrors are arranged to give a paraboloidal shape and reflect the incident solar radiation to a common point, i.e., the receiver.

Receiver

The receiver of a Scheffler dish is placed at the focus of the dish so as to capture the incident solar radiation and transfer it to the thermal medium used in the system. The critical component is generally a dome-type absorber fixed onto the structural steel frame. The receiver works on the thermo siphon principle in which cold water sinks down from header and hot water/steam moves up to the header. The receivers are designed according to application of end user and also the size of receiver depends on the size of the focus and storage requirement.

Tracking System

The tracking system enables the dish to be focused towards the sun to capture maximum possible direct radiation during the day. It also tracks the sun as it changes its position during the seasons. The tracking system

Table 9.2 Specifications of collector.

Parameter	Specification
Type	Solar grade mirror with low iron content or Aluminium reflector with glass like finish for an outdoor use. Aluminium reflectors should have a proper back support for the precise reflective surface area.
Material	Tempered and toughened Solar grade glass/Glass infused Aluminium/Aluminium tested for scratches and durability.
Shape	Flat mirror cut in required sizes for different models.
Size	All dimensions of a mirror should be less than the diameter of the receiver.
Thickness	2-3 mm for Solar Grade Mirror/0.5mm for Aluminium Reflector.
Reflectivity	At least 90%.
Reflective coating	High quality silver back coating.
Protective Coating/ Edge sealing (on cutting of mirrors for different Sizes)	Standard weather protection coat on all the sides after grinding the corners with the help of emery paper and application of appropriate coating on solar grade. Mirrors. Special weather protection coat to be applied for back side of mirrors to be used in the coastal and industrial region with high pollution levels.
Durability	At least 10 years for Solar Grade Mirrors and 10 years for Aluminium mirrors.

may include the following set of equipments: Electrical motor, Gearbox, Sprockets and chains, Counter weight with rope, Screw shaft, solar radiation sensor, Wind sensor and Timer.

Storage system

The thermal storage subsystem is a part of the circulation system. It extracts heat from the circulating fluid when the temperature becomes

Figure 9.5 Aluminium mirror.

Figure 9.6 Solar Grade mirror.

Table 9.3 Specifications of mirror.

Parameter	Specification
Material	Tempered and toughened Solar grade glass/Glass infused Aluminium/Aluminium tested for scratches and durability.
Shape	Flat mirror cut in required sizes for different models.
Size	All dimensions of a mirror should be less than the diameter of the receiver.
Thickness	2-3 mm for Solar Grade Mirror/0.5mm for Aluminium Reflector.
Reflective coating	High-quality silver back coating.
Reflectivity	At least 90%.
Protective Coating/ Edge sealing (on cutting of mirrors for different Sizes)	Standard weather protection coat on all the sides after grinding the corners with the help of emery paper and application of appropriate coating on solar grade mirrors. Special weather protection coat to be applied for back side of mirrors to be used in the coastal and industrial region with high pollution levels.
Durability	At least 10 years for Solar Grade Mirrors and 10 years for Aluminium mirrors.

too high. Likewise, when the temperature is too low, it supplies stored heat to the fluid.

Solar Tracking panel

This automatic tracking system for Scheffler Reflectors (16 m^2) sizes provides a comfortable solution for daily tracking of the primary reflector. It rotates the polar mounted axis of the dish and makes sure the focal point remains in the exact position throughout the day. In the evening, the dish is reversed automatically to morning position.

Sensor based tracking: The sensor detects the position of the sun and gives signals to the motor to rotate the reflector. The reflector starts to track automatically each morning as soon as direct sunlight passes on the sensor.

Table 9.4 Specifications of receiver.

Parameter	Specification
Design	Spherical Dome type receiver
Material	Boiler grade Mild Steel
Size & Thickness	Diameter is based on the thermic fluid flow rate required as per the process. Thickness is based on the pressure required in the system 350-400 mm diameter; 8 mm thick plate for dome-type Receiver.
Strength	Subjected to hydro test as per the applicable pressure rating of the system.
Durability	With normal use as per prescribed operation and maintenance manual, the receiver is designed to last 15 years.
Receiver Insulation	1. Insulating material with wire mesh and cladding covering the back and the side of the receiver. 2. Insulating Material – Light Resin Bonded (LRB) Mattress OR Rockwool OR Ceramic Wool Cladding. Material – Aluminium/Stainless steel 3. Insulation Thickness – 60 mm Cladding Thickness – 22 SWG 4. Durable up to system life of 10 years

Tracking operates when the sun shines. When there is no sun, no tracking is required.

Automatic reverse function: In the afternoon, the reflector hits a switch and reverses into the (east-facing) morning position where it will wait until the sun is in the east again (next morning). The time in the afternoon when the reflector starts turning back into its morning position can be set by moving the stopper for the reverse switch.

Motor and Gears: The motor and gears provide a sturdy slow movement of the reflector. The speed of the motor allows it to catch up with the position of the sun in case the sun comes out after a longer period of clouds during that day. Motor and gears are covered against rain.

Automatic and manual Mode: Scheffler dish rotation can be fixed on automatic or manual mode. This depends upon the operation of the plant.

Figure 9.7 Receiver insulation.

Figure 9.8 Receiver support structure.

Table 9.5 Specifications of receiver support structure.

Parameter	Specification
Design	Steel structure of standard IS sections and angles fixed on a cement concrete structure.
Material	Mild Steel structure and Cement Concrete.
Strength and Durability	Designed to withstand 160 Km/hr. of wind speed with a life span of 20 years.
Fixing of receiver on Support Structure	Fixed on the Mild Steel angle structure with help of nut bolts and positioned at focal point with receiver plane making an angle equal to latitude angle of the location with vertical.

Figure 9.9 Tracking system.

In the present project, four Scheffler disc collectors have been integrated to provide thermal energy for operation of the chiller during sunny hours of the day. The area of each Scheffler disc is 16 m^2 (16 square meter). A concentrating primary reflector (or Scheffler's collector) tracks the sun, focusing sunlight on a fixed receiver. The primary

Table 9.6 Specifications of tracking mechanism.

Parameter	Specification
Mechanism	Automatic mechanism (Microprocessor based/Timer based) for daily tracking and manual mechanism for seasonal tracking in 16m^2 Scheffler dish system Automatic mechanism (Timer-based) for daily tracking and Semi-Automatic mechanism with the help of switches for seasonal tracking in 32 m^2 scheffler dish system.
Features	Automatic reversal in the evening to bring it back to the morning position. Accuracy: +/- 0.2°
Weather protection	Standard weather protection box for all outdoor tracking components.
Strength and Durability	All the standard pieces of equipment related to the tracking system are designed to withstand 160 km/hr. of wind speed with a life span of 20 years.
Equipment used	• Standard Electric Motor, Rope with counter weight, Gear box, Chain, Timer and Threaded Bars for timer based daily tracking. • Set of photo sensors, chains, motors, gears and actuators for microprocessor based daily tracking. • Screw shafts on top and bottom for the seasonal tracking.

reflector produces a converging beam of sunlight aligned with an axis of rotation. The axis of rotation is parallel to the axis of the Earth. The clock mechanism rotates the primary reflector around its axis of rotation at a rate of one revolution per day. It keeps the reflected beam aligned with the axis of rotation as the sun moves. Each morning, the operator has to rotate the primary reflector back to a starting position in which the receiver is illuminated, and start the clockwork. After a few days the operator has to adjust the angle between the axis of rotation and the reflector to accommodate the seasonal variation in the height of the sun. Once the receiver is illuminated and the clockwork is in motion, the spot of focused light remains on the receiver throughout the day. The seasonal variation in the height of the sun requires changing not only the angle between the primary reflector and its axis of rotation, but also

Figure 9.10 Storage tank.

Figure 9.11 Storage support system.

Table 9.7 Specifications of Storage tank.

Parameter	Specification
Design	Dish End type Header Drum
Material	Boiler grade Mild Steel
Size and Thickness	• Diameter is based on the thermic fluid flow rate required as per the process under consideration • Thickness is based on the pressure required in the system
Storage Material	Hot Water/Steam
Strength and Durability	• Subjected to hydro test as per the applicable pressure rating of the system • With normal use as per prescribed operation and maintenance manual, it is designed to last for 20 years
Insulation	• LRB mattress with Aluminum cladding over the pipe • Insulation Thickness – depending on temperature changes • Cladding Thickness – 22 SWG • Should withstand up to system life of 10 years

Table 9.8 Specifications of Storage tank structure.

Parameter	Specification
Design	Steel structure of standard IS sections and angles fixed on a cement concrete structure
Material	Mild Steel structure
Protection from corrosion	Application of two layers of red oxide paint over cleaned bars and sections followed up by application of two layers of enamel paint OR Application of red oxide paint followed up by application of a layer of primer plus two layers of epoxy paint and one layer of PU coat on bars and sections cleaned with copper slag
Strength & Durability	Designed to withstand 160 Km/hr. of wind speed with a life span of 20 years

the shape of the reflector. The primary reflector comprises a number of flat glass mirror or acrylic mirror facets. The axis of daily rotation is located exactly in north-south direction, parallel to Earth's axis. It runs through the centre of gravity of the primary reflector. Thus the reflector always maintains its gravitational equilibrium and the mechanical tracking device (clock mechanism) doesn't need to be driven by much force to rotate the primary reflector. The inclined cut produces the typical elliptical shape of the Scheffler-Reflector. The focus is located on the axis of rotation to prevent it from moving when the reflector rotates. The focus stays fixed, which is very useful. The receiver doesn't have to be moved.

9.3.1.5 Vapor Absorption Machine (VAM)

Refrigeration is defined as the process of extracting heat from a low temperature medium and transferring it to a high-temperature heat sink. Refrigeration maintains the temperature of the heat source below that of

Figure 9.12 Scheffler dish collectors.

its surroundings while transferring the extracted heat to a heat sink. There are mainly two types of system used for refrigeration and air conditioning.

1. Vapor compression technology.
2. Vapor absorption technology.

The most commonly used refrigerant-absorbent pairs in commercial systems are:

1. Water-Lithium Bromide ($H2_0$-LiBr) system for above 0°C applications such as air conditioning. Here water is the refrigerant and lithium bromide is the absorbent.
2. Ammonia-Water (NH_3-$H2_0$) system for refrigeration applications with ammonia as refrigerant and water as absorber.

Vapor Compression System

The vapor compression refrigeration system consists of an evaporator, compressor, condenser and an expansion valve. The refrigeration effect is obtained in the cold region as heat is extracted by the vaporization of refrigerant in the evaporator. The refrigerant vapor from the evaporator is compressed in the compressor to a high pressure at which its saturation temperature is greater than the ambient or any other heat sink. Hence when the high-pressure, high-temperature refrigerant flows through the condenser, condensation of the vapor into liquid takes place by heat rejection to the heat sink. To complete the cycle, the high-pressure liquid is made to flow through an expansion valve. In the expansion valve the pressure and temperature of the refrigerant decrease. This low-pressure and low-temperature refrigerant vapor evaporates in the evaporator, taking heat from the cold region. It should be observed that the system operates on a closed cycle. The system requires input in the form of mechanical work. It extracts heat from a cold space and rejects heat to a high temperature heat sink.

In the vapor-compression refrigeration cycle, refrigerant enters the evaporator in the form of a cool, low-pressure mixture of liquid and vapor. Heat is transferred from the relatively warm air or water to the refrigerant, causing the liquid refrigerant to boil. The resulting vapor is then pumped from the evaporator by the compressor, which increases the pressure and temperature of the refrigerant vapor. The hot, high-pressure refrigerant vapor leaving the compressor enters the condenser where heat is transferred to ambient air or water at a lower temperature. Inside the condenser, the refrigerant vapor condenses into a liquid. This liquid refrigerant then flows to the expansion device, which creates a pressure drop that reduces the pressure of the refrigerant to that of the evaporator. At this low pressure, a small portion of the refrigerant boils (or flashes), cooling the remaining liquid refrigerant to the desired evaporator temperature. The cool mixture of liquid and vapor refrigerant travels to the evaporator to repeat the cycle.

Vapor Absorption Technology overview

Absorption systems use heat energy to produce a refrigerating effect. In these systems the refrigerant, i.e., ammonia, absorbs heat at a low temperature and pressure during evaporation and releases heat at a high temperature during condensation. A solution known as absorbent, water is used to absorb the vaporized refrigerant (after its evaporation at low pressure). This solution, containing the absorbed vapor is heated at a higher pressure. The refrigerant vaporizes and the solution is restored to its original concentration for recirculation. In a double effect absorption machine, the latent heat of condensation of the refrigerant generated in the first stage generator is used in a second-stage generator to enhance the efficiency of the cycle. When the refrigerant undergoes a series of evaporation, absorption, pressurization, vaporization, condensation, throttling, and expansion processes, absorbing heat from a low-temperature heat source and releasing it to a high-temperature sink, so that its state is restored to its original one, it is said to have completed a refrigerating cycle.

The main advantage of absorption chillers is their ability to utilize waste heat streams that would be otherwise discarded. In terms of energy performance, motor-driven vapor compression chillers will beat absorption chillers every time. Still there are specific applications where absorption chillers have a substantial advantage over motor-driven vapor compression chillers. Some of those applications include:

1. For facilities that use a lot of thermal energy for their processes, a large chunk of heat is usually discarded to the surrounding as waste. This waste heat can be converted to useful refrigeration by using a VAM.
2. For facilities that have a simultaneous need for heat and power (cogeneration system), absorption chillers can utilize the thermal energy to produce chilled water.
3. For facilities that have high electrical demand charges. Absorption chillers minimize or flatten the sharp spikes in

a building's electric load profile and can be used as part of a peak shaving strategy.

4. For facilities where the electrical supply is not robust, expensive, unreliable, or unavailable, it is easier to achieve heat input with a flame than with electricity. Absorption chillers uses very little electricity compared to an electric motor driven compression cycle chiller.
5. For facilities, where the cost of electricity versus fuel oil/gas tips the scale in favor of fuel/gas. Various studies indicate that the absorption chillers provide economic benefit in most geographical areas, due to the differential in the cost between gas and electric energy.
6. For facilities wanting to use a "natural" refrigerant and aspiring for LEED certification (Leadership in Energy and Environmental Design) absorption chillers are a good choice. Absorption chillers do not use CFCs or HCFCs – the compounds known for causing Ozone depletion.
7. For facilities implementing clean development mechanism (CDM) and accumulating carbon credits, the absorption use coupled to waste heat recovery and cogeneration system help reduce problems related to greenhouse effect from CO_2 emission.

   Vapor absorption system allows use of variable heat sources: directly using a gas burner, recovering waste heat in the form of hot water or low-pressure steam, or boiler-generated hot water or steam.

A vapor absorption chiller (VAM) is a machine to produce chilled water using heat source such as steam, hot water, gas, and oil. It seems unreasonable to achieve cooling with heat, but that is what happens inside an absorption chiller. These machines use only a small fraction of electricity as compared to the conventional vapor compression chillers. Vapor absorption systems work with non-CFC environmentally friendly refrigerants such as water or ammonia. The current project has been designed for ammonia-water absorption system because the temperature up to ---10 to 20°C can be achieved in the cold storage and avoids the freezing of water as ammonia is used as refrigerant. Unlike lithium-bromide system, these chillers operate at moderate pressures and no vacuum is required till – 30°C and these chillers do not suffer from the problem of crystallization.

The VAM used in the project has 15 kW capacity is modified and consists of:

1. Heat Source
2. Evaporator
3. Generator
4. Condenser
5. Heat exchanger
6. Rectifier
7. Absorber

VAM Components

Heat Source: Waste heat from the exhaust of engine, water jackets of a reciprocating engine, auxillary firing and solar thermal route is recovered for use through heat exchangers. The heat from the flue gas, thermal route and engine exhaust can be utilized to drive the hot water or steam to run chillers.

Refrigerant: Vapor absorption systems work with non-CFC refrigerants such as water or ammonia as refrigerant. These systems find acceptability in the commercial air conditioning or process cooling. The ammonia-based absorption system that uses ammonia as a refrigerant, finds use in low-temperature applications.

Cooling Water: Cooling water is required to cool the absorbent and the refrigerant vapor. The cooling water flows through the Absorber and the Condenser items of the VAM. The heat gain in the cooling water is rejected in a cooling tower and thus is an open loop. Around 2% of water is lost as a result of evaporation, drift and blow down in the cooling tower that needs make-up.

Generator: The purpose of the generator in the system is to generate the refrigerant (ammonia) through the system. It accomplishes this by separating the ammonia (refrigerant) from the water (absorber). In the generator, a high-temperature energy source, flows through tubes that are immersed in a dilute solution of refrigerant and absorbent. The solution absorbs heat from the hot water, causing the refrigerant to boil and separate from the absorbent solution. As the refrigerant is boiled away, the absorbent solution becomes more concentrated. the concentrated absorbent solution goes to the absorber after cooling and the refrigerant ammonia migrates to the condenser.

Condenser: The purpose of the condenser is to condense the refrigerant vapors to liquid inside the condenser; cooling water flows through tubes and hot refrigerant vapor fills the surrounding space. A heat transfers from the refrigerant vapor to the water, refrigerant condenses on the tube surfaces. The condensed liquid refrigerant is collected in the bottom of the condenser before travelling to the expansion device. The cooling water system is typically connected to a cooling tower. Generally, the generator and condenser are inside the same shell.

Expansion Device: From the condenser, the liquid refrigerant flows through an expansion device into the evaporator. The expansion device is used to maintain the pressure difference between the high-pressure (condenser) and low-pressure (evaporator) sides of the refrigeration system by creating a liquid seal that separates the high-pressure and low-pressure sides of the cycle. As the high-pressure liquid refrigerant flows through the expansion device, it causes a pressure drop that reduces the refrigerant pressure to that of the evaporator. This pressure reduction causes a small portion of the liquid refrigerant to boil off, cooling the remaining refrigerant to the desired evaporator temperature. The cooled mixture of liquid and vapor refrigerant then flows into the evaporator.

Evaporator: The evaporator and absorber are contained inside the same shell, and this allows refrigerant vapors generated in the evaporator to migrate continuously to the absorber. In the evaporator, the liquid ammonia is converted to the vapor because of low pressure, and till the latent heat is taken by the ammonia vapor because of the said process, cooling effect is obtained in the evaporator. Evaporation system is kept inside the Cold storage system and a fan coil unit is used for the evacuation of cool from evaporator tube to the cold storage system.

Absorber: Inside the absorber, the refrigerant vapor is absorbed by the water. As the refrigerant vapor is absorbed and converted to liquid (i.e., formation of aquous-ammonia solution). In the absorber, concentrated absorber (water) is spread after cooling. The cool absorber attracts the vapor ammonia and is converted to liquid aqua-ammonia solution. Because of this phenomena, low pressure inside the evaporator and absorber is maintained. However, a absorption exothermic reaction takes place. Therefore heat removal inside the absorber is required to maintain the absorption property, so the condensed water is also supplied to the absorber for removal of exothermic heat.

Heat Exchanger: The liquid aqua-ammonia solution is continuously pumped from absorber to the generator through the heat exchanger. In the heat exchanger, the aqueous ammonia solution takes heat from the strong solution of absorber.

Rectifier: The rectifier is a sort of heat exchanger cooled by the water, which is also used for cooling the condenser. Due to cooling the remaining water vapor mixed with the ammonia refrigerant also gets condensed along with some particles of ammonia. This weak solution of water and ammonia drains down to the analyzer and then to the generator.

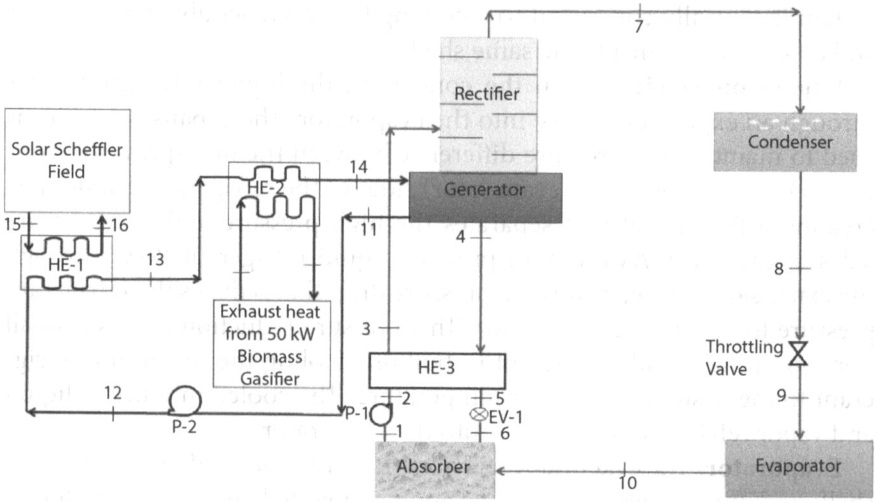

Figure 9.13 Line diagram of Solar biomass hybrid model for cold storage.

Pump: The strong solution of ammonia and water is pumped by the pump at high pressure to the generator.

Efficiency of Vapor absorption Machine (VAM)
Efficiencies of absorption chillers is described in terms of Coefficient of Performance (COP) and is defined as the refrigeration effect divided by the net heat output,

$$COP = \frac{\text{Cooling capacity obtained at evaporator}}{\text{Heat Input for the generator}}.$$

9.3.1.6 *Cold Storage Unit*

Cold storages are meant to preserve the perishable commodities of food items for a longer period with retention of the original colour, flavor and taste. Cold storage of fruits, vegetables and milk products adds an important value to these products and these products can be used with nearly the same flavor and nutritional value. Once harvested, fruits and vegetables must be stored under proper conditions, the most important of which are temperature and humidity. Each fruit or vegetable has its own ideal set of conditions at which it can be stored successfully for a longer period of time.

In addition to proper temperature and humidity, all fruits and vegetables must be kept in a dark, aerated environment. Produce must not be allowed to freeze and should be protected from animal pests. The goal of storage is to keep them in a dormant state. One other note, fruits and vegetables should always be stored separately. Fruits release ethylene, which speeds the ripening process of vegetables. Fruits are also very susceptible to picking up the taste of nearby vegetables.

Factors affecting Cold storage design and Construction:

During the design and construction of cold storage, the following precautions should be taken in order to avoid excessive and abnormal stresses.

Wind pressure (During Erection): Wind forces can impose excessive stress on an incomplete insulation system and any necessary temporary supports and propping shall be provided. Particular attention was paid to the possibility of abnormal stresses being applied to panel clips during panel erection.

Wind pressure (Completed Structure): Wind forces imposed on the complete insulation envelope will vary with the locality of the store and the position of any dominant openings. Exposed panels and their fixing systems have been designed to resist wind pressure.

Figure 9.14 Cold storage unit (outside view).

Dead and Imposed Loads: The structure, suspended insulating ceiling and load-bearing walls shall be capable of supporting the maximum combination of all loads.

The size of cold storage is 5 X 4 X 3 (m). The total volume accounts to 60 m^3. The cold storage has 20 tons vegetable storing capacity. The panel insulation for the cold storage designed for the project has pre-fabricated outer facing with an injected polyurethane core (PUF) and the insulant used in cold storage is polyurethane foam. The facing material used in the project is plastic coated galvanized steel sheeting. In order to achieve maximum cold room insulation, the same insulated panels that are used for walls and floor are used for the ceiling as well. Through fan oil unit, evacuated inside the cold storage. All latent heat of ammonia taken as cool in cold storage.

The biomass is fed in a downdraft gasifier and due to the process of gasification, the producer gas is produced inside the gasifier. This producer gas is fed to the engine-generator unit which is a producer gas-based engine. The conversion of electricity/power generation takes place inside the engine-generator unit and is fed to the smart grid for end use of consumers. The engine exhaust which would have been a waste in normal

Figure 9.15 Cold storage (inside view).

gasification process is used to run VAM by using it in heat recovery system. Thus the utilization of waste energy takes place inside the heat recovery system and is an efficient example of energy management. In case of peak load, the energy exhaust is sufficient to run the VAM; however, in case the peak load isn't obtained from the engine or the engine is idle, a provision of auxiliary firing from burner has been made to supplement the heat from from engine exhaust. The third provision of running VAM is through hot water cycle by using Solar Sheffler disc units for solar thermal generation.

In case of non-sunny days, the VAM will operate on engine exhaust and/or auxiliary firing and in case engine exhaust isn't available, the system will run on auxiliary firing only. Thus the system is designed in such a way that the cold storage can run for 24 hours (if needed) on solar, auxiliary firing and engine exhaust. It can also run on the combination of either of the two or on the combination of all the three as per load requirement. The VAM data is observed in the data logger, which was installed in the experimental site. The data logger gave the solar field inlet and outlet temperature, pressure, expansion valve pressure, and flash vessel temperature, and this data was obtained per minute. The data was then analyzed on an hour basis taking the average values of all components into consideration.

9.4 Performance Evaluation

During Daytime: When the amount of solar radiation is good, i.e., during daytime, the biomass gasifier is not operational. The heat is obtained through Scheffler dishes. Each Scheffler dish has a receiver where heat is concentrated and is absorbed by water which is inside the pipe at certain pressure connected with receiver. The heated water then passes through heat recovery unit gaining more heat and is fed to Vapor Absorption Machine which provides cooling to cold storage. Actually the water transfers its heat to Generator of Vapor Absorption Machine and water after losing heat is circulated back to solar field to gain heat.

During nighttime: When there is no radiation/solar energy then the Biomass Gasifier is operational. The producer gas obtained from the gasifier is fed in internal combustion engine and we obtain 50 kW of mechanical energy, which is further coupled with the generator to produce electricity; also the exhaust gas coming out from internal combustion engine is passed through the Heat Recovery unit with the help of pipe and the exhaust are losses it heat and, atmosphere through chimney, thus from Heat Recovery unit the water gains heat and is fed to Vapor Absorption Machine and cooling is obtained.

During Low Radiation: At times there are conditions when there is neither high radiation nor low radiation so at such circumstances biomass gasifier and Scheffler dish are both used. The producer gas obtained from the gasifier is fed to internal combustion engine where energy is obtained and the exhaust gas coming out from internal combustion engine is passed through the Heat Recovery unit with the help of pipe. Some amount of producer gas is also directly fed to Heat Recovery unit where the producer gas is burned to release heat and also the exhaust gases lose heat and the waste gases which don't have more energy is passed out. On the other hand heated water from Scheffler dishes is also passed through Heat Recovery Unit and its energy increases which is passed to Vapor Absorption Machine.

9.4.1 Thermodynamic Analysis

The first law of thermodynamics is used to analyse the performance evaluation of hybrid cold storage using solar and exhaust heat of biomass gasifier. The energy equation is mainly defined for the major components of the hybrid cold storage using solar and exhaust heat of biomass gasifier. The energy analysis has been investigated by several authors. The following energy balance equation of major components of cooling system are expressed as:

The useful heat gain (Q_1) is defined as

$$Q_1 = Q_{solar} \times \eta_{E,Scheffeler} \tag{9.1}$$

Where Q_{solar} is the solar energy falling on Scheffler field may be expressed as

$$Q_{solar} = A_{ap} \times DNI \tag{9.2}$$

The energy efficiency ($\eta_{E,Scheffeler}$) of the Scheffler field is expressed as

$$\eta_{E,Scheffeler} = (\eta_{opt} \times COS\theta) - a_1 \frac{(T_m - T_o)}{DNI} - a_2 \left[\frac{(T_m - T_o)^2}{DNI} \right] \tag{9.3}$$

Where η_{opt} is optical efficiency of the Scheffler technology, a_1 is the first order coefficient of the collector efficiency ($W/m^2\text{-}^0C$), a_2 is the second order coefficient of the collector efficiency ($W/m^2\text{-}^0C^2$), T_m is the mean temperature of the fluid is defined as

$$T_m = \frac{T_{15} + T_{16}}{2} \tag{9.4}$$

The mass flow rate of fluid (m_{fluid}) is calculated as

$$m_{fluid} = \frac{\eta_{E,Scheffeler} \times A_{ap} \times DNI}{\Delta T \times C_p} \tag{9.5}$$

Where, C_p is the specific heat of heat transfer fluid, ΔT is the temperature difference of heat transfer fluid across the Scheffler field is expresses as

$$\Delta T = (T_{15} - T_{16}) \tag{9.6}$$

$$\text{Generator: } Q_g = m_4 \cdot h_4 + m_7 \cdot h_7 - m_3 \cdot h_3 \tag{9.7}$$

$$\text{Condenser: } Q_C = m_7 \cdot h_7 - m_8 \cdot h_8 \tag{9.8}$$

$$\text{Evaporator: } Q_e = m_9 \cdot h_9 - m_{10} \cdot h_{10} \tag{9.9}$$

$$\text{Pump-1: } W_{p-1} = m_3 \cdot h_3 - m_2 \cdot h_2 \tag{9.10}$$

$$\text{HE-2: } Q_{HE-3} = m_2 \cdot h_2 + m_4 - h_4 - m_3 \cdot h_3 - m_5 \cdot h_5 \tag{9.11}$$

The coefficient of performance of the ammonia-based cooling system is expressed as

$$COP = \frac{Q_e}{(Q_g + W_{P-1})} \tag{9.12}$$

9.5 Results and Discussion

The system was run continuously from 10 a.m. to 6 p.m. for 8 hours from 2013-2015 on Solar thermal energy (during day) and on non-solar days auxiliary firing was used as offset to solar thermal energy; the system worked on biomass gasifier in night and the power was fed to smart grid for end consumption as electricity. The system was run continuously in different conditions of optimal load, under load and overload conditions to evaluate its performance evaluation and the following results were obtained.

The variation of COP with Generator temperature is shown in Figure 9.16. The results showed the COP to increase with the increase in generator temperature. However, the maximum COP of 0.60 was obtained at generator temperature of 85.91 °C.

The results of various calculations showed the trend of COP at various operating condenser temperatures. The COP of the system (VAM) increased with the increase in Condenser Temperature. The highest condenser temperature obtained was 26.83 °C at COP of 0.60.

The minimum evaporator temperature of - 4.24 °C was achieved. The result of calculations shows the trend at lower evaporator temperatures greater COP is achieved.

Auxiliary firing quantity was inversely proportional to exhaust output/electrical load consumption. The relationship between auxiliary firing required to run the VAM and power generation, corresponding to engine exhaust is given. In case of peak output, i.e.. if the energy output of generator is 50 kw, no biomass firing is required as the engine exhaust is sufficient to run the VAM. However, in case our electric output is zero. i.e., generator is not producing any electricity/output, in that case at least 18.66 kg of biomass is required to run the VAM, as the engine exhaust will be zero. Similarly, if the electric output is 47.093 (i.e., less than desired Output),

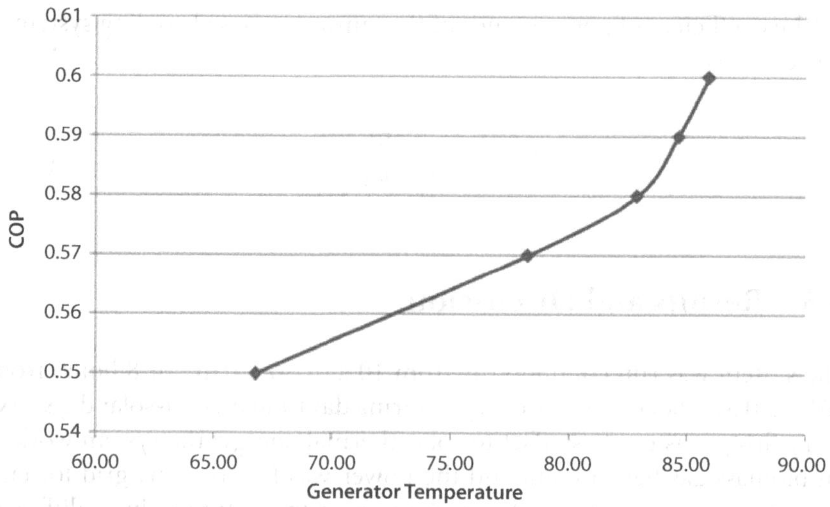

Figure 9.16 Effect of COP on generator temperature.

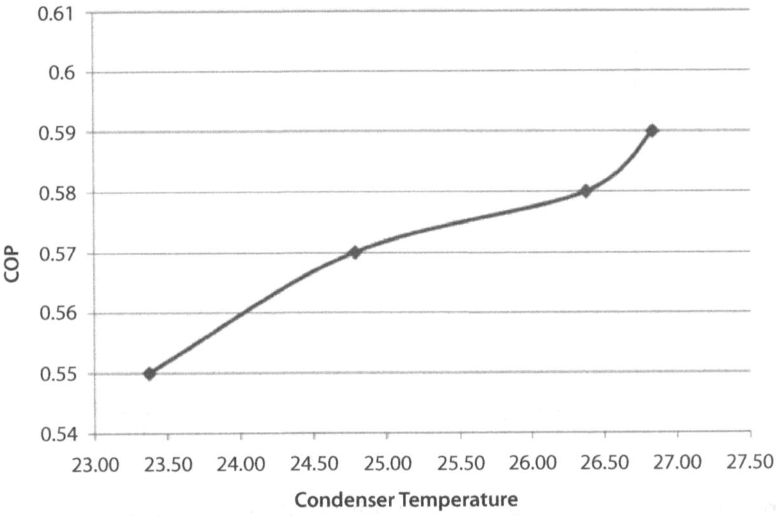

Figure 9.17 Effect of condenser temperature on COP.

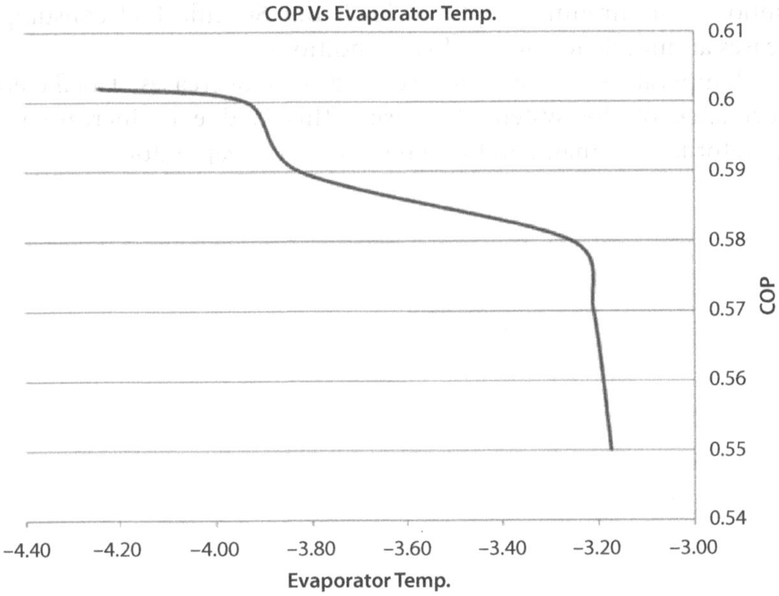

Figure 9.18 Effect of evaporator temperature on COP.

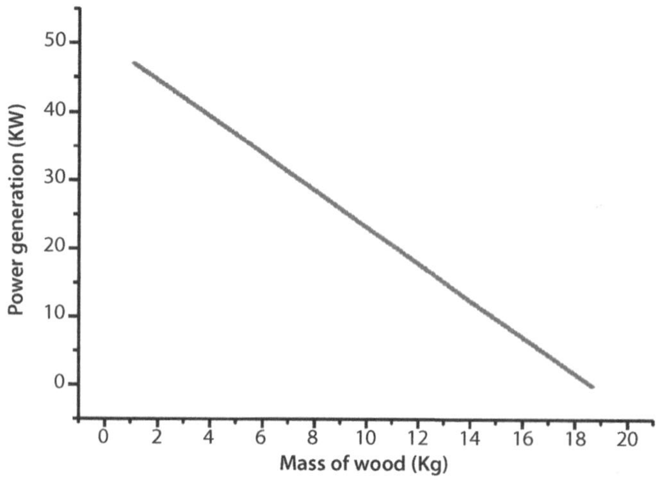

Figure 9.19 Mass of Wood used in auxiliary firing v/s Power generation.

nearly 1.085 kg of biomass is required for auxiliary firing. This shows for the running of VAM, auxiliary firing is dependent on engine exhaust, which is in turn dependent on the power generation.

The graph 9.20 shows the relationship between specific biomass fuel consumption with optimal load. The curve shows the biomass fuel consumption is minimum in optimal load. But Specific fuel consumption increases at underload and overload conditions.

As Temperature required for cooling area decreases, Coefficient of Performance of the system decreases. This is due to increase in the Evaporator load to maintain low temperature of evaporator.

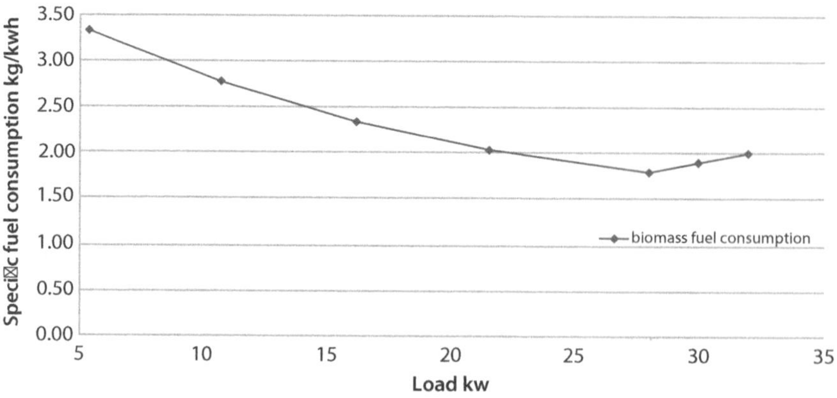

Figure 9.20 Specific fuel consumption vs. load.

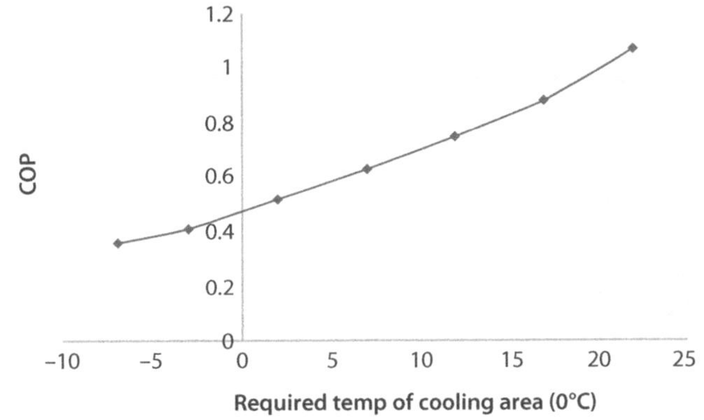

Figure 9.21 Variation of COP with required cooling area temp.

Experiment shows the result of varying the condenser pressure. As system pressure goes up or down beyond optimum pressure of 11.48 bar, the COP decreases. This happens because refrigeration effect produced by ammonia decreases.

The above graph shows that if electrical load is minimum, then exhaust heat is not sufficient to operate the VAM, in case auxiliary firing is required. If the system operates at optimal load then exhaust is sufficient to operate the VAM.

It shows that exhaust heat increases with load. If electrical load is low, and then exhaust heat is not sufficient to operate the VAM, in case auxiliary firing is required.

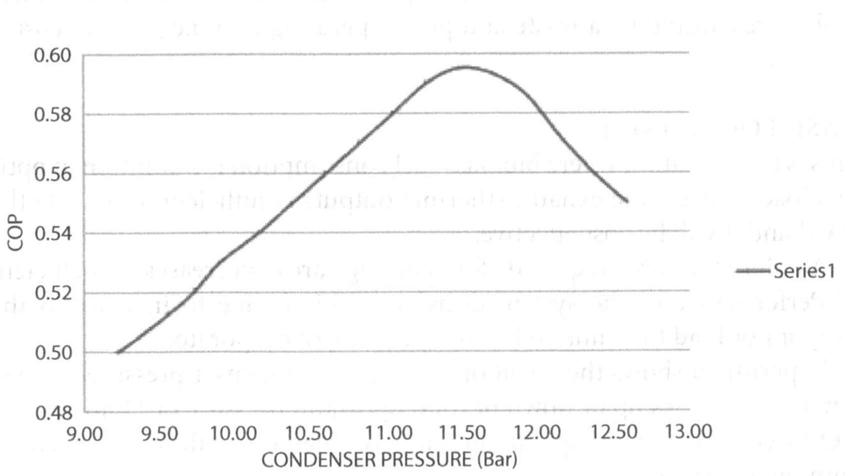

Figure 9.22 Effect of COP with Condenser pressure (bar).

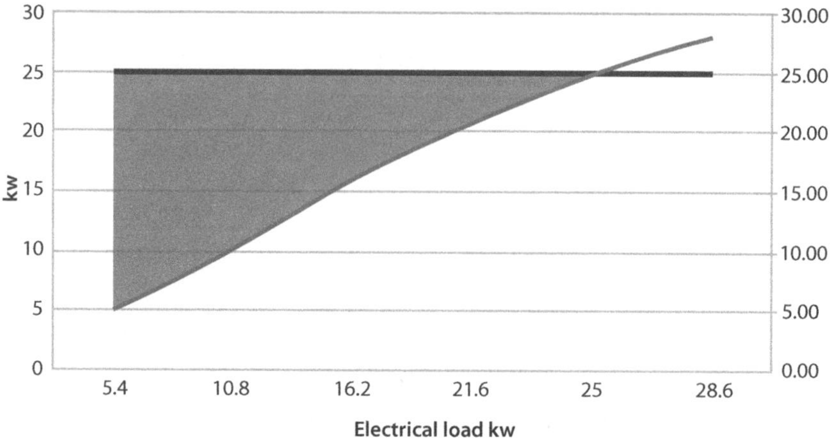

Figure 9.23 Variation of exhaust heat vs. electrical load.

CASE 1 Underload
If plant runs in this condition, Specific fuel consumption increases at underload as enhanced engine exhaust is not sufficient to operate the VAM, then auxiliary firing is required, so specific fuel consumption increases and also energy cost increase.

CASE 2 Overload
If plant runs in this condition, Specific fuel consumption will also increase, and Gas engine exhaust (thermal output) will be excess, then extra thermal energy would be a waste and plant operating cost, i.e., overall cost is increased.

CASE 3 Optimal load
This is the condition where biomass fuel consumption is minimum in optimal load. Gas engine exhaust (thermal output) is sufficient to operate the VAM and it will be cost effective.

As Temperature required for cooling area decreases, Coefficient of Performance of the system decreases. This is due to increase in the Evaporator load to maintain low temperature of evaporator.

Experiment shows the result of varying the condenser pressure. As system pressure goes up or down beyond optimum pressure of 11.48 bar, the COP decreases. This happens because refrigeration effect produced by ammonia decreases.

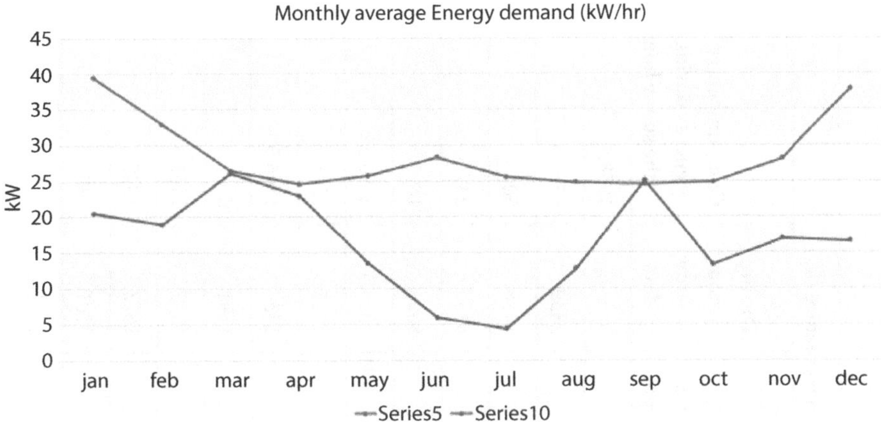

Figure 9.24 Monthly energy demand (KW/hr) to run vapor absorption machine.

The energy demand in winter gets increased because ambient temperature is very low and thus heat losses are more. So in winter energy demand for such type of cold storage under specifications is more as compared to summer, where ambient temperature is optimal for its use.

The data analyzed for the whole year reveals that during the months of March and September nearly zero or much less biomass was required as complementary energy substitute to run the cold storage as required energy was available to run the system through solar Scheffler technology only. However, complementary biomass firing was required for the winter months as solar Scheffler technology was not able to run the system alone as less energy was available through solar thermal technology.

The above result shows that only during the months of March and September did the system work effectively on solar energy only. It also means that the required energy used for cold stoarge was available mostly in the months of March and September. The data for each day for each month was accessed and it was found that Solar Scheffler technology was efficent for other months than May, June and July. It is because of the reason that the solar incident angle modifier is incresed in summer, i.e., reflector to receiver radiation is not at a proper angle, and it is because of the seasonal losses that in the months of June and July, Scheffler technology is not effective to run cold storage independently.

The above data reveals that if all other accessories and common systems/parts of a cold storage are having same cost, a vapor compression machine (VCM) which works on conventional energy, i.e., electric load of nearly same

Table 9.9 Energy consumption and availability to run cold storage.

Month	Energy required for cold storage (kWh/ month)	Energy from solar (kWh/ month)	Energy from biomass (kWh)			Biomass reauire (kg)		
			During night	During non sunny sunny hours	Complimentary energy during during the sunny hours	During night	During non sunny days	Complimentary energy during the sunny hours
Jan	29340	4246	14670	6507	3917	0	3253.5	1958.5
Feb	22138	3678	11069	4678	2713	0	2339	1356.5
Mar	19677	6883	9839	2856	100	0	1428	50
Apr	17671	7113	8836	1227	495	0	613.5	247.5
May	19124	4160	9562	1619	3783	0	809.5	1891.5
Jun	20368	1576	10184	2631	5977	0	1315.5	2998.5
Jul	19013	1034	9506	3450	5022	0	1725	2511
Aug	18446	3254	9223	2876	3093	0	1438	1546.5
Sep	17679	6589	8840	2382	0	0	1191	0
Oct	18472	3315	9236	3054	2867	0	1527	1433.5
Nov	20250	4088	10125	3347	2690	0	1673.5	1345
Dec	28138	3447	14069	6202	4419	0	3101	2209.5
Total	250316	49384	125158	40829	35076	0	20414.5	17538

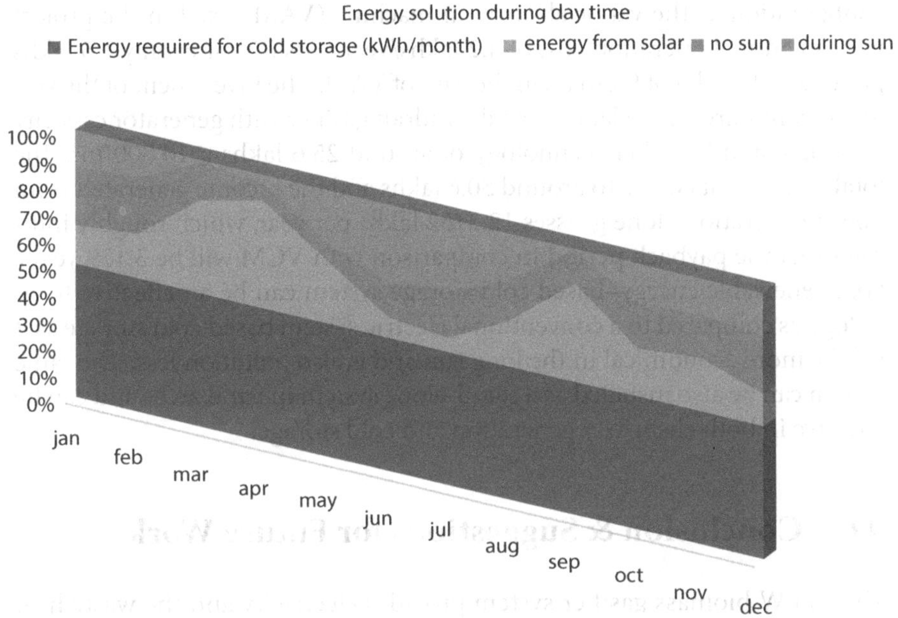

Figure 9.25 Energy optimization during daytime.

Table 9.10 Economic analysis of the system.

Technology	**Rs Lakh/year (annual operation cost)**
VCM with electric (2.5 COP)	4.2 at the rate Rs8/unit
VAM with biomass gasifier and solar	1.5 at the rate rs 4/kg
Income from power generation	13.3152 Lakh/year
Investment	**Rs**
Biomass gasifier with genset	2500000
Sheffier dish	2560000 at the rate RslOOOO/ m2
Total	5060000
Pay back in comparison with VCM	3.159498476 year

configuration as the vapor absorption machine (VAM) used in the project with an annual operation cost of 4.2 lakhs/year @Rs 8/unit is only 1.5 lakhs per year @ Rs 4/kg of biomass in the case of VAM. The investment of the system is configured as 25 lakhs for a downdraft gasifier with generator capacity of 50 kW and Scheffler technology of around 25.6 lakhs @10,000/m^2. The total investment comes to around 50.6 lakhs and the income generated from power generation alone grosses 12.3152 lakhs per year, which roughly indicates that the payback period in comparison with VCM will be 3.159 years. So, a renewable energy–based cold storage system can be an effective technology as compared to a conventional electric system based cold storage as it will be more economical in the long run and is also pollution less. The same system can be also installed as a stand-alone system in rural areas and can be effective in both electricity generation and cold storage.

9.6 Conclusion & Suggestions for Future Work

The 50 kW biomass gasifier system provides electricity and the waste heat of the engine (exhaust) is used as the main source of energy for the cold storage. During the daytime when the sunshine is available the Scheffler dishes along with engine exhaust provide heat to the system. During evening, when the sun is not available and the gasifier is run to meet the evening electricity load, the cold storage will operate only on engine exhaust. In case no sunshine is available and engine exhaust is also not sufficient to meet heat requirement of VAM for cold storage, provision has been made for firing of producer gas in the Heat Recovery Unit for uninterrupted heat supply to VAM. Nearly 3.2 MWhr of electricity generated by the plant has been fed to a resistive load of 40 kW during the year.

Suggestions for Future Work

A lot of further investigations need to be done in solar-Biomass hybrid models for power generation and cold storage. Some of the suggestions on which future research can be done are:

1. The system needs to be tested in different climatic conditions and topographies.
2. In the given system, ammonia was used as refrigerant and water as absorber in VAM. Further optimization of absorber and refrigerant is possible, and such refrigerants and absorbers need to be cheap, reliable and efficient.

3. The COP of present system was observed 0.60. There is enough scope for COP improvement.
4. The material used in Cold storage needs to be tested in different conditions and a different material may be tested and used which is cost effective, reliable and moisture proof.
5. Normal experiments and observations can be done in gasifier design to improve the overall efficiency of the system.
6. The Scheffler disc collector used in the project has enough scope for improvement.
7. The system needs to be run for a longer period of time and the performance evaluation can be done for a longer period of time in advance research programmes.
8. The system needs to be optimized further and more Scheffler discs can be added so that the system works effectively on solar thermal energy.
9. Automatic tracking of receiver can be done so that cosine losses and incident losses are minimized.
10. The system can be hybridized with other renewable energy sources like PV systems and another CST technologies.

References

Conklin JC, Szybist JP. 2010. A highly efficient six-stroke internal combustion engine cycle with water injection for in-cylinder exhaust heat recovery. *Energy*, 35:1658–64.

Jiu Huang, Klaus Gerhard Schmid and Zhengfu Bian, Removal and Conversion of Tar in Syngas from Woody Biomass Gasification for Power Utilization Using Catalytic Hydrocracking. *Energies*. 2011, 4, 1163-1177; doi:10.3390/en4081163.

Kalyan K. Srinivasan, Pedro J. Mago and Sundar R. Krishnan 2010. Analysis of exhaust waste heat recovery from a dual fuel low temperature combustion engine using an Organic Rankine Cycle. *Energy*, vol. 35, issue 6, 2387-2399.

Kumar R.C, A. Sonthalia and R. Goel. 2011, Experimental study on waste heat recovery from an international combustion engine using thermoelectric technology. *Journal of Thermal Sciences*,Vol 15, no. 4, pp. 1011-1022.

Papadopoulo A.M, S. Oxizidis and N. Kyriakis 2010. Perspective of solar cooling in view of the development in the air-conditioning sector. *Renewable and Sustainable Energy Reviews* vol. 14, issue 1, pp. 312-322.

Rajiv Varshney, J. L. Bhagoria, C. R. Mehta. Small scale biomass gasification-an overview. *Journal of Engineering, Science and Management Education*, Vol. 3, 2010, 33-40.

Roudriguez H.M.C, A. Rodriguez, P.A. Lecuona and Ventas R. 2010. Flat plate thermal solar collector efficiency: Transient behavior under working conditions. Model description and experimental validation. *Renewable and Sustainable Energy Reviews*, vol. 14, issue 1, pp. 122-131.

Sathiamurthi. P. 2011, Design and Development of Waste Heat Recovery System for air Conditioning Unit, *European Journal of Scientific Research*, Vol. 54, No. 1, pp. 102-110.

Wang T, Zhang Y, Zhang J, Shu G, Peng Z. 2012. Analysis of recoverable exhaust-energy from a light-duty gasoline engine. *Applied Thermal Engineering*, 53(2):414–419.

Worley, M, and Yale, J. *Biomass Gasification Technology Assessment: Consolidated Report*. United States: NREL Report, 2012. Web. doi:10.2172/1059145.

Xiaodong Zhang, K.T. Chau, 2011. An automotive thermoelectric–photovoltaic hybrid energysystem using maximum power point tracking. *Energy Conversion and Management* 52(1):641-647.

Younes Chhiti, Mohammed Kemiha, 2013. Thermal Conversion of Biomass, Pyrolysisand Gasification: A Review. *International Journal of Engineering and Science*, Vol. 2 (3), 75-85.

Index

Also of Interest

Check out these other related titles from Scrivener Publishing

By the Same Author/Editor:

Progress in Solar Energy Technology and Applications, edited by Umakanta Sahoo, ISBN 9781119555605. This first volume in the new groundbreaking series, Advances in Renewable Energy, covers the latest concepts, trends, techniques, processes, and materials in solar energy, focusing on the state-of-the-art for the field and written by a group of world-renowned experts. *NOW AVAILABLE!*

A Polygeneration Process Concept for Hybrid Solar and Biomass Power Plants: Simulation, Modeling, and Optimization, by Umakanta Sahoo, ISBN 9781119536093. This is the most comprehensive and in-depth study of the theory and practical applications of a new and groundbreaking method for the energy industry to "go green" with renewable and alternative energy sources. *NOW AVAILABLE!*

Other Related Titles:

Energy Storage 2nd Edition, by Ralph Zito and Haleh Ardibili, ISBN 9781119083597. A revision of the groundbreaking study of methods for storing energy on a massive scale to be used in wind, solar, and other renewable energy systems. *NOW AVAILABLE!*

Nuclear Power: Policies, Practices, and the Future, by Darryl Siemer, ISBN 9781119657781. Written from an engineer's perspective, this is a treatise on the state of nuclear power today, its benefits, and its future, focusing on both policy and technological issues. *NOW AVAILABLE!*

Zero-*Waste Engineering 2nd Edition: A New Era of Sustainable Technology Development*, by M. M. Kahn and M. R. Islam, ISBN 9781119184898. This book outlines how to develop zero-waste engineering following natural pathways that are truly sustainable using methods that have been developed for sustainability, such as solar air conditioning, natural desalination, green building, chemical-free biofuel, fuel cells, scientifically renewable energy, and new mathematical and economic models. *NOW AVAILABLE!*

Sustainable Energy Pricing, by Gary Zatzman, ISBN 9780470901632. In this controversial new volume, the author explores a new science of energy pricing and how it can be done in a way that is sustainable for the world's economy and environment. *NOW AVAILABLE!*

Sustainable Resource Development, by Gary Zatzman, ISBN 9781118290392. Taking a new, fresh look at how the energy industry and we, as a planet, are developing our energy resources, this book looks at what is right and wrong about energy resource development. This book aids engineers and scientists in achieving a true sustainability in this field, both from an economic and environmental perspective. *NOW AVAILABLE!*

The *Greening of Petroleum Operations*, by M. R. Islam et al., ISBN 9780470625903. The state of the art in petroleum operations, from a "green" perspective. *NOW AVAILABLE!*

Emergency Response Management for Offshore Oil Spills, by Nicholas P. Cheremisinoff, PhD, and Anton Davletshin, ISBN 9780470927120. The first book to examine the Deepwater Horizon disaster and offer processes for safety and environmental protection. *NOW AVAILABLE!*

Biogas Production, Edited by Ackmez Mudhoo, ISBN 9781118062852. This volume covers the most cutting-edge pretreatment processes being used and studied today for the production of biogas during anaerobic digestion processes using different feedstocks, in the most efficient and economical methods possible. *NOW AVAILABLE!*

Bioremediation and Sustainability: Research and Applications, Edited by Romeela Mohee and Ackmez Mudhoo, ISBN 9781118062845. Bioremediation and Sustainability is an up-to-date and comprehensive treatment of research and applications for some of the most important low-cost, "green," emerging technologies in chemical and environmental engineering. *NOW AVAILABLE!*

Green Chemistry and Environmental Remediation, Edited by Rashmi Sanghi and Vandana Singh, ISBN 9780470943083. Presents high quality research papers as well as in depth review articles on the new emerging green face of multidimensional environmental chemistry. *NOW AVAILABLE!*

Bioremediation of Petroleum and Petroleum Products, by James Speight and Karuna Arjoon, ISBN 9780470938492. With petroleum-related spills, explosions, and health issues in the headlines almost every day, the issue of remediation of petroleum and petroleum products is taking on increasing importance, for the survival of our environment, our planet, and our future. This book is the first of its kind to explore this difficult issue from an engineering and scientific point of view and offer solutions and reasonable courses of action. *NOW AVAILABLE!*

Printed in the USA/Agawam, MA
September 16, 2026

910661.005